How Life Began

How Life Began

Evolution's Three Geneses

Alexandre Meinesz

Translated by Daniel Simberloff

The University of Chicago Press
Chicago and London

Alexandre Meinesz is professor at the University of Nice–Sophia Antipolis. He is the author of more than 200 articles and several books, including *Killer Algae*, also published by the University of Chicago Press.

Daniel Simberloff is the Nancy Gore Hunger Professor of Environmental Studies at the University of Tennessee. He has written several hundred articles and coedited *Strangers in Paradise: Impact and Management of Nonindigenous species in Florida* (Island Press, 1997).

The University of Chicago Press, Chicago 60637
The University of Chicago Press, Ltd., London

Printed in the United States of America

17 16 15 14 13 12 11 10 09 08   1 2 3 4 5

ISBN-13: 978-0-226-51931-9 (cloth)
ISBN-10: 0-226-51931-7 (cloth)

Library of Congress Cataloging-in-Publication Data

Meinesz, Alexandre.
How life began : evolution's three geneses / Alexandre Meinesz ; translated by Daniel Simberloff.
p. cm.
Includes bibliographical references and index.
ISBN-13: 978-0-226-51931-9 (cloth : alk. paper)
ISBN-10: 0-226-51931-7 (cloth : alk. paper) 1. Life—Origin. I. Title.
QH325.M36 2008
576.8′3—dc22
2007035929

⊗ The paper used in this publication meets the minimum requirements of the American National Standard for Information Sciences—Permanence of Paper for Printed Library Materials, ANSI Z39.48-1992.

CONTENTS

ACKNOWLEDGMENTS

It was in 1998 that a minibus picked me up in front of a hotel in Seattle and took me and the other participants to an ecology seminar high up in the wooded hills of Washington State. I knew none of the others, and I found myself sitting between two senior American researchers who were discussing what seemed to me, a European, to be an odd subject. They were comparing their scores for the number of countries, states, and counties they had visited during their university careers. I kept quiet and tallied in my mind my feeble score compared to those of these great travelers. I was even more surprised at their arguments regarding the ranking, debated with real passion, of certain countries or counties that were difficult to access or were out of the way for these great conference goers. One of these scientific globetrotters was Daniel Simberloff. It was only after my lecture that Dan came to see me and, in remarkably good French, began to express interest in my research and my publications. That was the beginning of a long and amiable collaboration. He painstakingly translated my first book describing my battle to gain recognition of the harmfulness of an introduced invasive alga. It is the publication of this present work that marks the high point of our collaboration. Many people have played a part in putting the final touches to this book and have been involved in the long process of preparation, but all these friends are aware that this work could not have seen the light of day without the commitment, enthusiasm, and unfailing help of Daniel Simberloff, who translated the text. It is to him that I owe the greatest debt of gratitude.

I would like, too, to mention my daughter, Marjorie Meinesz, to whom I am indebted for the faithful and meticulous presentation of my drawings that illustrate the book. Finally, I would also like to acknowledge in particular the work of my editor, Christie Henry, who encouraged me unstintingly and was able to bring the work to fruition thanks to the perspicacity of the excellent reviewers, who remain anonymous to me, whom she selected, and to whom I am also very grateful.

INTRODUCTION

The painting is small and dark, but famous. It is one of the 30 or so known paintings of Johannes Vermeer. The subject is unusual: an astronomer contemplating a globe of the heavens. The astronomer is almost certainly Antoni van Leeuwenhoek, a scientist who lived in Delft, Holland, in the seventeenth century. At various times he worked in the cloth trade, measured out wine, and administered estates; he was a bailiff, surveyor, and astronomer. But he is best known for perfecting and using an excellent small microscope. He was the first to see microscopic life, organisms not visible to the naked eye. For the first time, a man could see microorganisms swarming in water, spermatozoa in sperm, red blood cells. At a stroke, life took on a new dimension.

Since then, many of the mysteries of life and its origins have been elucidated with increasingly abstract dimensions. The accumulation of discoveries is dizzying. Each specialist delves more deeply into his well of expert knowledge but never sees the bottom. In the process, he forgets much of what lies above and surrounds his specialty. This is why popular media tell us, in disordered, intermittent fashion, about multiple discoveries made in the diverse disciplines of the life sciences. But the connections between these bits of knowledge and their relative importance are blurred in our minds.

I have tried to see this tableau clearly, and Vermeer's painting has helped.

If you visit the Louvre in Paris, go and admire the gentle face of Antoni van Leeuwenhoek looking at the representation of the heavens on a sphere. With one hand, he tightly grasps a table, the material objects of this world.

With the other hand, he brushes the starry globe, a synthesis of scientific data about our universe. In the background, a canvas hangs on a wall. Experts have identified the subject: Moses saved from the waters of the Nile by the pharaoh's daughter.

Everything is in its place: man firmly attached to the concrete aspects of his existence, the religious and spiritual phenomena so prominent then, and the remarkable scientific knowledge that was accumulating so quickly.

What did Van Leeuwenhoek want when he commissioned and sat for this work? And what was Vermeer trying to suggest? In some of his paintings, Vermeer, the Catholic convert, attested to his faith. It is therefore probable that for him, the globe representing the universe was allegorical. His model, the astronomer, seemed to want to attain the celestial infinitude, but with hesitation. He gives the impression of reflecting on the spiritual guide located somewhere in the heavens.

Van Leeuwenhoek, the pragmatic Protestant, eagerly embraced all the paths to knowledge that opened in his century. He earned a diploma in surveying and dreamed about the infinite heavens while looking up at stars captured on a celestial sphere. Vermeer therefore painted him as an astronomer. Van Leeuwenhoek lived close to those who traced the routes of ships of the Dutch East India Company—routes that led to unknown lands and gave access to exotic spices. Vermeer did a second portrait of him in which he is surrounded by measuring tools in front of continents and oceans depicted on maps: this is *The Geographer* displayed today in the Städel Museum in Frankfurt, Germany. Van Leeuwenhoek was posing for Vermeer before he had perfected his microscope. But perhaps he already was thinking about exploring the infinitesimal, given his skill in using optical lenses. Another Dutch painter immortalized him with a microscope in his hands and a globe of the Earth in the background.

Faced with such a profusion of discoveries and wealth of knowledge to explore, Antoni van Leeuwenhoek surely posed the questions of the Greek philosophers: Where did man come from? Where did life originate? What is life?

Nowadays, another question is urgent: Where are we taking life?

Thus, before the unknown of tomorrow, we must better integrate and relate the formidable current hypotheses about the history of life, only recently divested of the religious beliefs and superstitions that stymied or biased this effort for so long. This research about the truth of our distant past is also useful in demystifying many ill-advised proposals embracing recent scientific advances.

This essay is aimed at a large public with elementary knowledge of biology. It covers the origin of the major stages in the elaboration of life. It deals with the domains of biology, paleontology, and ecology as well as philosophy and theology.

It synthesizes many scientific data gleaned, for the most part, quite recently (within the past fifteen years). I present this synthesis in nine chapters illustrated by personal experiences or events in my professional life as researcher, teacher, and environmental manager. The anecdote about my admiration for the paintings by Vermeer that depict Antoni van Leeuwenhoek, discoverer of microbial life, is one of these vignettes that allows a less academic treatment of a subject that is usually hard to communicate.

The essence of this book is to elucidate three origins, or geneses, all governed by a major evolutionary tendency captured by the motto "unity is strength." I emphasize four types of chance that have governed the construction of life. Some of these claims are novel.

After evoking the emergence of man and his consciousness (chapter 1), the geneses I depict are those of the first bacteria (chapter 2), the first animal and plant cells (chapter 4), and multicellular organisms (chapter 7). This work underscores the preeminence of bacteria (chapter 3) and unicellular animals and plants (chapter 6) in the biosphere and in current biodiversity. It develops an understanding of the dominant position of these lineages in which evolution occurs in the interior of isolated cells. Finally, I split evolution into four types of fortuitous events (random occurrences or contingencies) that have sculpted life. Three are creative (genetic mutations, genetic recombination that occurs in sexual reproduction, and natural selection, which are covered in chapter 5); the fourth is destructive (the great geological cataclysms, discussed in chapter 8).

The penultimate chapter (9) stresses how hard it is to perceive the immense time scale of evolution and relate it to the cyclic time scale of the natural changes observed within one generation. The epilogue (chapter 10) points to the many ways to interpret the *grandeur of life*. I emphasize the key role of a structuring principle that has given rise to each of the three geneses of life: the principle of union. It was the association and union of organic molecules, especially RNA and DNA chains, that produced the first bacteria. It was the addition and union of bacteria that gave rise to the genealogical lineages of animals and plants. And finally, it was the union of cells of the same species that allowed the growth of multicellular organisms visible to the naked eye. The grandeur of life therefore results from a major evolutionary tendency based on sociality and solidarity.

Union has produced the force of life; thanks to union, life has flourished. The union of men in civilizations and the uniting of their knowledge is similarly at the base of our greatness. Only our cohesion and solidarity will allow us to solve the environmental problems that we will confront in the decades and centuries to come. The epilogue accentuates the necessity for a great leap in public engagement, which should aid humanity in uniting to take greater responsibility for the fate of life on Earth, and which will require a better education about the environment and all that pertains to life.

As an ecologist and specialist in marine habitats and the first organisms that colonized Earth (algae), I offer different insights than those of microbiologists, geneticists, or paleontologists who typically popularize the main themes developed in this book.

Like Van Leeuwenhoek, I rejoice in trying to understand many types of knowledge. To share them in this book, I have forced myself to use jargon-free language to aid the lay public in understanding the fabulous history of life. In the notes and references, those who are curious can find explanations, elaborations, and the sources of my synthesis.

Though science has, as its classic image, that of a mischievous Einstein sticking out his tongue, I keep in my heart that of the gentle face of Antoni van Leeuwenhoek and his prudent gesture. His head throbbed with remarkable facts and he surely had a troubled mind. I invite you to follow him: stay in tune with the facts of your daily life, but whether you are a believer or an atheist, discover the recent marvelous discoveries on the origins of life. They will make you reflect and dream.

CHAPTER 1

Henri's Cave

On Our Origin

My friend Henri was an insurance agent and then an investment advisor; for many years he has managed people's finances. He inherited uncultivated arid land in Ardèche, his home department in France. One evening in January 1995, during a meeting of our Lions Club, he told us with great emotion about the discovery under one of his properties of an immense cave with walls covered by dozens of prehistoric paintings. Several days later, the press reported the news. Three excited spelunkers had made the wonderful discovery that now bears the name of one of them: Chauvet Cave. The frescoes of the cave have gradually been revealed, and Henri, waxing enthusiastic, asks us repeatedly, "Have you seen it? It's under my property; it's my cave!"

Henri was privileged to visit "his" cave with specialists in Paleolithic art. He was flabbergasted. He showed us photos and described the surprisingly realistic and elaborate frescoes. The gifted prehistoric artists—Cro-Magnons—had mastered the art of painting murals, interweaving lions, rhinoceroses, bears, mammoths, horses, and bison.[1] A first fantasy is even modestly depicted behind a rock outcrop: two legs and female genitals.

As in many other caves, the artists "signed" their works with many negative and positive imprints of their hands. Charcoal, remains of torches used to light the cave, and pigments date to about 32,000 years ago. These paintings are currently the oldest known in the world.

Henri has read up on the people of this period in our distant past. He wanted to know everything about the ancient history of his property. At the club, he told us about the likely customs of the first inhabitants of Ardèche. As he tells his stories, we can see that he has had a revelation.

Thanks to this discovery, Henri has traveled excitedly on the obscure pathways of the history of our origin. Party to a remarkable discovery, he has been transformed, surprised to have learned so much about wall painting. Henri can speak more confidently about it than anyone else; it is his cave, and he has been able to contemplate the prehistoric frescoes with his own eyes. Respected for his new competence, he enjoys telling about his first-hand experience.

Moreover, his expertise in commercial matters has led him as a matter of course to reflect on the exploitation of this unexpected treasure. This is why, when France decided to take his cave from him—to annul his inheritance of the land and works of his distant ancestors, to steal part of his dreams—he saw red. Henri had to defend his interests in court, because the state had expropriated his property for a ludicrous sum. His land and therefore his cave had been valued at less than five cents per square meter.[2] Henri, the proud Ardèche native, scion of a long line of local squires, owner of a château topped by an impressive eleventh-century dungeon, was forced to concentrate all his energies on getting his property properly valued. He felt as if he had been taken for a naive peasant from the wilds of Ardèche who had been scornfully granted a pittance in exchange for his patrimony. His words were not sugar-coated when he described his legal battle against these churls (several high-ranking Paris bureaucrats), whom he cursed endlessly.

As for his rock paintings, recurring questions arose about their meaning. Hypotheses abound. The most widely held today refers to ancestral rites perpetuated by a caste of initiated and respected men (or women?): the shamans. They were simultaneously the scientists, artists, priests, sorcerers, and druids of earliest history. It is a gripping subject, appealing to the imagination. Henri, promoted to the fraternity of "experts in Paleolithic rites," frequently explains his hypotheses. This has become a ritual; after "Hello, Henri," we always add, "And your cave, any news?"

The Spiritual References of the Beginning of Life

We all have our own ideas about how life appeared on Earth and the stages of its evolution. It is an exciting topic, taught in universities and often addressed by the popular media. Far from being settled by immutable facts, it is a perpetually evolving subject. And relevant knowledge has expanded considerably in the last few decades.

For Christians and Jews, the reference is a text that appears in the first book of both the Bible and the Torah, entitled Genesis (from the Greek word "genesis," meaning origin or birth). For Muslims, different sūras of the Koran similarly describe the creation of the universe in several days. These religious descriptions of the different stages in the creation of life can be summed up as follows. In the beginning God created the heavens, the Earth, and the light that separates them from a shadowy chaos. On the second day He divided the waters of the heavens (rain) from those of the earth (the oceans). On the third day God created plants, and on the fourth day, the sun, moon, and stars. On the fifth day He created birds and marine animals, and on the sixth day, terrestrial animals, men, and women.[3] The date of this dazzling creation of life on Earth has been fixed between 3671 BC (according to the Jewish calendar) and 4004 BC, at noon on October 23 (according to detailed historical and biblical research undertaken by an Anglican archbishop in the seventeenth century).[4]

Asian myths and Taoist and Confucian religions attest to a profound attachment to nature; for those who follow these beliefs, nature is sacred. Its genesis is believed to have been the doing of a giant, P'an Ku, who was himself created from chaos by the yin and the yang (the two opposing complementary forces). Though there are many versions of his story, they usually agree on the different stages in the creation of the elements of earth and of life. They report that P'an Ku worked throughout his life to give the Earth a friendlier appearance. In an egg, in which he grew for 18,000 years, he first separated heavens from earth, then created mountains and dug river channels. But one day, exhausted by his labors, the giant collapsed and died. From his remains, five sacred mountains arose. His dying breath became the wind and clouds, his last cry thunder. His flesh became soil, his blood was transformed into rivers, and his bones formed rocks. The sweat of this giant turned into rain so profuse that its accumulation produced seas and oceans. His hair rooted to produce plants, while his entrails metamorphosed into serpents and quadrupeds. Finally, the parasites that infested his body gave rise to the human species. Shintoists believe that god initially created the first man (Izanagi) and first woman (Izanami). They, in turn, created the world as we know it, beginning with the seas and an island (the main island of Japan). They then created the other Japanese islands and different gods—one for each environmental element (mountains, rivers, fire, and so on)—and, after various events, the gods of the sun, the night, and the sea. For Shintoists, a god resides in each natural element.

The Scientific History of the Beginning of Life

These mystical tales of the creation of life have a scientific analog, elaborated over the last two centuries by virtue of paleontology and the discoveries of the laws of evolution, heredity, and ecology. This version can be summarized as follows. Life appeared on Earth between 3.8 and 3.5 billion years ago, and the fossils of the first forms of life are from bacteria. Unicellular organisms, ancestors of animals and plants, arose approximately 2.7 billion years ago. Finally, animals and plants composed of many cells proliferated beginning only 570 million years ago. So there were several stages in this origin of life, several geneses. Life as we know it today developed by slow evolutionary pathways broken by a succession of cataclysms (impacts of large meteors, bouts of intense volcanism, periods of severe glaciation) that favored certain kinds of survivors, which subsequently multiplied, dispersed, and diversified. This is how species with complex anatomies (including *Homo sapiens*) arose from simpler living forms.

Our Recent Ancestors

In addition to these religious and scientific accounts, which are greatly simplified, many other ideas bear on our evolution and anchor us in the deepest roots of life. Traces of our ancestors evoke emotions. They make us dream about their lives, which were not so very long ago. Curious, we want to know more about our predecessors in the human lineage—such facts concern our blood, our features, our genetic identity. The Mormons, by ideology, expend remarkable efforts to find as many official documents as possible about our ancestors. Their files on many interlaced genealogies go back to the Middle Ages.

We are also moved by the many superb vestiges of ancient civilizations whose manners and customs have been reconstructed.

As for Cro-Magnon man, Neanderthal man, and the ancestral apes, the traces are scarcer, the various branch points less evident. Questions arise: Where is the true cradle of humanity? In what period did a highly evolved ancestral form of ape become a prehistoric human? What is the difference between a humanoid ape and an archaic human? An increasing number of expeditions seek remnants of our distant ancestors. Many paleontologists now rummage through their habitations and exhume their remains. The more of them they find—and the number of sites is immense—the more they seem lost in a dense bush of branches of ancestral great apes, ape-men, or prehumans that cohabited or succeeded one another on Earth.

In a book full of amusing drawings, Stephen J. Gould (1941–2002), a biologist, geologist, and paleontologist at Harvard University, called into question the two classic images that sum up our evolution—that of the chimpanzee metamorphosing into man and that of the genealogical tree of our species in the form of an oak in which man represents the key branch, the highest and strongest.[5]

The imaginary cartoon in which the chimpanzee by stages stands up, loses its hair, and assumes more human-like traits does not reflect reality. We are not descended from chimpanzees; rather, chimpanzees and humans arose from two different lineages with a single common base—a sort of ancestral ape.

The image of a dense bush with irregular ramifications in all directions is more realistic than that of a symmetric, immense oak in depicting the profuse lines of ancestral apes and hominids that coexisted or succeeded one another. Our true genealogical tree has many dead or broken branches corresponding to extinct apes and prehumans. If we want to indicate the population size of each species by the thickness of the branches, we end up with a bush that does not conform to normal plant development, because it has both robust branches that suddenly dwindle to twigs (corresponding to the rare survivors of some formerly dominant species) and minuscule twigs that thicken into veritable trunks higher up (representing dominant species today, such as our own). It is at the extremities of these branches, which all reach the same height, that the last descendants are located; we are among them, along with other present-day apes, including chimpanzees.

The origin of apes is harder to conceive. However, it has been established that the first apes arose from small mammals. The skeleton of the earliest known placental mammal, dated to 125 million years ago, resembles that of a rodent.[6] Similarly, biologists have deciphered the stages of life that allow us to retrace the transformations leading from the humble, archaic, unicellular animal to humans.

Among all the stages of these metamorphoses of life, the one most cited is the transformation from ancestral ape to man. On television, leading specialists regularly present our ancient ancestors realistically and excitedly. Among about twenty identified lineages,[7] four are mentioned most often:

- *Homo sapiens*—"the man who knows"—is our direct ancestor. Ancient representatives include Cro-Magnon man (Cro-Magnon is a hamlet near the French village of Les-Eyzies-de-Tayac in Dordogne where the first remains of modern man were exhumed).

- *Homo neanderthalensis*—"man of Neanderthal"—is another tribe that cohabited for tens of thousands of years with Cro-Magnon man. The first remains were exhumed in Germany in the Neander Valley in the basin of the Ruhr River, near Düsseldorf.
- *Homo erectus*—the definitely upright, vertical man. He walked upright on two legs, so his hands acquired other functions. He was a hunter and the first real voyager, venturing out of Africa, the presumed cradle of humanity.
- *Homo habilis*—the oldest "skilled man"—he is the first human to have worked with stones, 2.6–2.5 million years ago. This is the first lineage to have left the forest and spread into plains and marshes.

Our genealogical tree begins to become more confused if we look further back among the prehumans, where we find *Australopithecus* (which means apes [*pithecus*] from the south [*austral*]), well represented by the media star "Lucy," a large part of whose fossilized skeleton was exhumed and whose discoverers[8] presented her in 1974 as the ancestor of all humanity. Since then, many other remains of older humanoids have been found, each find pushing back the dawn of our emergence.

While Henri was getting excited by the artists of "his" cave, several discoveries occurred in Africa. The more fossil hominid remains that were unearthed, the more we found traces of great antiquity and the further back the date of our origin was pushed. The most ancient vestiges allow us to place the "birth" of our relatives between six and seven million years ago.[9] But, at the beginning of the twenty-first century, these indicators are still scarce. The number of nearly complete skeletons from the initial period can be counted on the fingers of one hand. This is why a simple fossilized femur, hard as rock, a tooth, or a fragment of a hominid cranium more than two million years old brings honor and fame to the discoverers. It is therefore necessary to extrapolate, interpret, and deduce behaviors or morphological traits from a tooth or a bone.[10] This research leads to logical conclusions. The further back we seek the origins of mankind, the more the traces are blurred and the greater the seeming resemblance to apes.

It appears that there was never a single birth episode at one time and place. All the facts we have gleaned about our most distant past demonstrate that a distinct species of man arose from a humanoid ape by small, successive changes. There was not a single, brutal metamorphosis but an evolution with some periods of accelerated change of traits and other periods—long ones—with little change.

The principles of ecology, ethology (the study of behavior), and genetics allow a cogent explanation of the transformations observed in fossils. Everything was probably set off by the emergence of nomadic behavior; prehumans wandered as families or tribes, first in Africa, where they left forests to explore savannas and marshes before traveling to other continents. For these hunters and gatherers, space was unlimited. But the environment was hostile and humans were not dominant—they were prey for many carnivores. Searches for new resources, flight to escape predators or the ferocious competition from enemy tribes: these behaviors all led to movement and isolation of small groups. Each isolated family or tribe adapted to new environments—forest, savanna, or marsh—and some genetic information, carrying a trait advantageous for life in certain favored habitats, was transmitted with increasing frequency to later generations. In fact, in small groups isolated for long periods, useful or neutral changes in genetic information are less likely to be "erased" or "leveled" by genetic recombination (the mixing of genetic information from two parents that occurs during sexual reproduction) than in large populations.

This dispersal led to the formation of different hominid varieties or species with distinct characteristics. Diverse types of prehumans arose from these diasporas, living in different regions and acquiring specific customs. We descend from one of these tribes.

Accepting Our Animal Origins: An Immense Spiritual, Philosophical, and Scientific Upheaval

To enrich our discussions with Henri at the club, I read a lot on this subject. But I am a biologist, not a historian specializing in paleoanthropology. So I have a personal perspective on this natural history. It seems to me that the evolution from ancestral ape to man seems is evident and that the proofs of the anatomical transformations, revealed by paleontological remains, are only descriptive details illustrating logical steps. Moreover, the proofs of our common ancestry with apes have been abundantly confirmed by comparative analyses of genetic information of humans and modern primates.[11] By contrast, the chronology of acquisition of knowledge about this subject has greatly impressed me.

The discovery of the relationship between apes and men was an earth-shattering event and has generated extraordinary controversy. In fact, just two centuries ago, in the Western world, it was necessary to agree—and to teach—that God created man and woman on the sixth day, about 4,000 years BC, and that all animals and plants are *creatures* (that is, *created* organisms), end of story. This dogma was religious, the religion was the state, and it was impossible to question it.

We should remember Galileo, who had the gall in 1632 to question a religious interpretation of the universe in which the biblical Earth was the fixed center around which the sun and stars revolved.[12] Threatened with torture, he was judged a heretic and forced to forswear his scientific convictions. He had to resign himself to recant his hypothesis in order to survive, groveling before his ecclesiastical judges, "*Eppur si muove*": but it does move! (This legendary anecdote was in fact proposed more than a hundred years after the death of Galileo by an Italian scrivener, Giuseppe Baretti.)

There were even excommunications following the first proclamations of evidence considered to be profoundly anti-biblical: the evolutionary relationship of apes and men, evolution of simple organisms into more complex ones, and the date of the first appearance of life on Earth, much earlier than the 4,000 BC suggested by the Bible. Some people spoke in terms of metaphors and tales, but they were often punished anyway. Others clearly affirmed what caused them to be tortured and burned at the stake by religious fundamentalists. Science has had its martyrs. The precursors and defenders of theories of evolution are most prominent among them, and they merit a bit more consideration.[13]

In fact, the French Revolution allowed three pieces of evidence of evolution to be recognized. For the first time in a large European country, religion was separated from the state. Therefore, beginning in 1800, the Frenchman Jean-Baptiste Pierre Antoine de Monet, Chevalier de Lamarck, professor at the Museum of Natural History in Paris and inventor of the word *biology*, had the audacity to publicize his theory of the transformation of species.[14] Using the examples of the evolution of winged creatures and of mollusks found fossilized in layers of rock, he explained clearly that his theory of the slow transformation of species could also apply to the transition from ancestral ape to man. He also demonstrated that life on Earth did not date back only the few millennia stated by the Bible. Lamarck estimated the age of the Earth at several million years. The mechanism that drove the process of evolution, according to Lamarck, was the transmission to their offspring of traits that organisms acquired during their lifetimes. He believed that acquired characteristics were inherited. His demonstration that transformation occurred was correct, but not his notion of the role of acquired characteristics.

The freedom arising from the French Revolution was short-lived. Beginning in 1809, the clergy were again the controlling power and Lamarck was more discreet. Dogmas and traditions got the best of him; Lamarck died discredited.

Nearly sixty years after the publication of Lamarck's theory, a creative English naturalist, Alfred R. Wallace (1823–1913), conceived of a mechanism of evolution more elaborate than the inheritance of acquired characteristics. He established a new concept that explained changes of species through time. Wallace was an explorer and naturalist, and he financed his research by capturing rare birds in distant lands (the Indonesian islands and Amazonian forests) and selling them to British zoos.[15] Pursuing birds in Malaysia, Indonesia, and New Guinea, he was struck by the differences between representatives of certain genera of birds and mammals, isolated from one another on nearby islands. He pointed to the examples of cassowaries (birds related to ostriches with a large, blue crest on their heads) and kangaroos; these animals are represented by different species in New Guinea, a mountainous island, and Australia, an isolated continent dominated by savanna with few mountains. A strait a mere 100 kilometers wide separates these two insular regions. Wallace interpreted these divergences as having resulted from natural selection following the immigration of groups of cassowaries and kangaroos from one island to the other. He understood that it is nature that favors (selects) individuals who have, by good fortune, inherited the traits best adapted to the new environment, and not, as Lamarck thought, the individual who has managed to adapt himself to the new environment and transmits to his offspring the adaptations he has contrived. Thus, different populations of a species tend to differentiate themselves indefinitely from the original, ancestral population by a process of natural selection.

The dispersal from a population of small groups of individuals with morphological characteristics that appear homogeneous but have different genetic bases constitutes the most common initial stage leading to the transformation of species. Novel morphologies (or other characteristics, such as physiological or behavioral traits) spread like waves from a central source. In isolated populations, confronting new environments and often in a condition of overpopulation and competition, natural selection of the best-adapted individuals occurs. This sequence is still considered the fundamental mechanism of evolution.

Wallace wanted to publish his theory. Toward this end, in 1857, he sent his observations to Charles Darwin (whom he considered his intellectual master) in the form of a manuscript mailed to England from the Indonesian jungles. Now, Darwin had twenty years earlier arrived at the same mechanism for the transformation of species. It had slowly matured in his mind and he had begun to draft a big book on the subject. He was beginning to be recognized for his reflections on the transformation of species through time. He based his ideas largely on his own observations, especially those on the finches of the Galapagos Islands, which he visited during a voyage around the world from 1831 to 1836. Of course, he was shattered to learn that Wallace, an unknown, was ready to rob him of the recognition for his great discovery. He discussed the problem with his colleagues. So due credit would be given to both authors, passages from their writings were transmitted jointly to the scientific community at a meeting in London in 1858. Two years after having received Wallace's essay, Darwin finally published his concept of the transformation of species by natural selection in a remarkable book in which he explained many original observations that supported the theory. Darwin was fortunate to live in the heart of one of the dominant nations of the era; he knew book publishers and scientific journal editors, he was surrounded by devoted, important friends, and he was a good communicator. His book, *On the Origin of Species,* was a sensation upon its publication in November 1859.[16] The theory of evolution by natural selection was widely diffused and supported by many authors who bore the brunt of religious criticism. The theory provoked a raging debate that increased Darwin's fame and, in certain circles, his notoriety. However, this first work on the origin of species treated only the question of the mechanism of the transformation of plants, insects, and birds by natural selection and domesticated animals by artificial selection by breeders (for pigeons, dogs, and horses). It was only 12 years later (and 70 years after Lamarck's theory) that Darwin finally applied the theory of evolution to humans, explaining how man descended from ancestral apes.[17] This was in 1871, and it was the main scientific revolution of the nineteenth century.

The highest official of the Anglican church in Great Britain mocked Darwin's colleagues in public, caustically asking them if it was through their grandfather or grandmother that they claimed descent from an ape. Others exclaimed, "My God, let us pray that this idea does not become known." Pope Pius IX (1792–1878) denounced the aberrations of Darwinism. An unfortunate Dominican monk, who had the impertinence to rally to the Darwinian cause, was summoned to Rome, where he was forced to forswear his earlier statements (as Galileo had been), and all his writings were banned.[18] Bishops condemned Darwin and his works. They were shocking. They rejected centuries of religious instruction. The Catholic and Anglican churches, reacting against this scientific modernism, did not hesitate to equate Darwinism with atheism. This was 1895, a little more than a century ago; two or three generations separate us from that state of mind. My father was then 7 years old.

In the twenty-first century, disciples of the creation as it is presented in Genesis, called creationists, have persisted in their way of describing the origin of life. They have succeeded in imposing their vision for 70 years in some colleges and secondary schools in conservative regions of the United States. The last law imposing the teaching of creationist theories, supported by President Ronald Reagan, was not abolished until 1987. Even today, a century and a half after the appearance of Darwin's theories, there are still many influential adherents of this conservative religious doctrine, which is very prominent in some fundamentalist Protestant churches centered in the United States.[19]

A Continuing Debate: The Appearance of Our Self-Consciousness

These episodes in the slow acceptance of scientific evidence allow us to understand the passion surrounding this subject. Clearly, the history of our roots, of the frail and hesitant basis of humanity, has strong philosophical and religious implications. The distinction between man (endowed with a highly developed sense of self-awareness) and animal does not reside solely in biology. Many specialists (biologists, geneticists, neurologists, paleontologists, anthropologists, philosophers, theologians, psychoanalysts, artists, journalists, novelists, and filmmakers) have opinions on this subject and can pique the interest of readers, listeners, and viewers for better or worse. The most minor news on the subject—a fossilized tibia dating from a crucial period—can rekindle the debate. The endless clues reported by thousands of detectives rummaging in our distant past allow the popular media to depict the dreams or reality of the extraordinary "transmutation" of animal into human. And everyone always wants to know more, which inspires others to look further for traces that will sustain the dreams.

This is why the last few decades have seen increasing introspection on the phenomenon of the humanization of ancestral ape into man.[20] It has been established that this evolution was accompanied by an increase in brain size. The addition and hereditary transmission of little supplementary RAM sticks of a particular type of memory would have sufficed to distinguish us from other ancestral apes or prehuman tribes. Nature proved this change to be an advantage, because it allowed man greater understanding and thereby allowed him to survive and reproduce better. This hereditary improvement permitted a cultural evolution as well, because the extra memory also served to transmit acquired knowledge better. Such knowledge accumulated from generation to generation.

Two simultaneous and complementary evolutions amplify the advantage. We can imagine the following process:[21] an ancestral ape, walking upright with his hands freed from locomotory functions, with a slightly better memory than his peers, noticed that, by throwing a stone, he was able to surprise a prey animal and kill it. He remembered this lesson, profited from it, and taught others in his tribe to do the same thing. He and his tribe ate better, which allowed them to produce more children than their neighbors and their distant cousins. The tribe of initiates therefore passed along the genes that caused them to have a better memory, and passed them in higher frequency than the genes of other individuals. In addition to the genetic transmission of this advantage, they taught their descendants the knowledge they had acquired. Sons and daughters subsequently passed along both the genetic and cultural advantages, generation after generation.

The transmission of acquired information multiplied the genetic advantage tenfold. This transmission is achieved by gestures (in *Australopithecus,* the hands were already free for six million years), by words, by painting and sculpture, by writing (which arose only 5,300 years ago), by printing (which appeared first in China at the beginning of the tenth century, then around 1450 in Europe), and today by light and sound waves and by the Internet.

Several different human tribes gradually evolved a more developed brain, allowing them to learn and to transmit their experiences. The human tribes best able to remember experiences and most assiduous at teaching them were "sapiens" (literally, "who knows"). These were the Cro-Magnon. They quickly dominated and spread on all continents. This is the tribe of our ancestors.

When, where, and for how long did the Cro-Magnon have the special, extra memory allowing them to improve on the rudimentary animal instinct to teach their young? It would be exciting to be able to go back in time to describe the historical stages in the transmission of acquired knowledge from generation to generation and in the emergence of self-consciousness. Biologists are convinced that knowledge and self-consciousness developed together. Neanderthal man probably acquired a substantial level of self-consciousness, and scientists have noted in current primates and many other animals glimmers of self-consciousness.[22] But we will never find fossils of self-consciousness or of the thoughts of the first representatives of the human race. The process of the emergence of self-consciousness in the past will remain abstract and indistinct. The same is true for the acquisition of the soul, the immaterial and immortal spiritual principle in which our personal identity resides. For the great majority of religions, the soul is the mind or breath of God, both immortal and separable from the body.

All of these concepts very quickly fascinated the first philosophers and religious figures who had the revelation of the enigmatic nature of the mind and of self-consciousness. These abstract and mythic notions led practitioners of certain precursors of monotheistic religions to establish a beginning, an origin. For them it is God who created the first two souls when he created Adam and Eve. The Bible says that they were created from dust, that their bodies and souls are the image of God. They are thus supernatural. Adam and Eve therefore have no parents; some perceptive artists have represented them without navels (or, out of modesty, have made grape leaves hide the sexual organs and the area where the navel would be found [plate 1]).[23] Present-day knowledge would surely have led the authors of Genesis to reserve for God alone the impulse to create the soul. In fact, although today we can:

- reconcile paleontological and genetic evidence of the transformation of ancestral ape into man (we have known since 1975 that 99 percent of proteins and nucleic acid sequences are identical for humans and chimpanzees[24]);
- sketch out the many contingencies of the mechanisms of evolution that have led a simple isolated cell to become a mammal; and
- point to great progress in the study of the neurological linkages that allow us to preserve information in our brains, leading to the acquisition of knowledge and self-consciousness,

we still remain wholly in the realm of the spiritual in explaining the process of acquiring a soul.

It is therefore logical to propose rewriting the first chapter of the Bible. The new Genesis would establish that Adam and Eve lived a very long time ago (why not 4004 BC, as the Bible indicates?). They would have navels, because they were produced by similar humans. They would nevertheless remain noteworthy, because they would be distinguished from their contemporaries in having received the breath of God. Ever since, that acquisition has been transmitted to all their descendants. God would thus have given a soul to man—not a body.

The French theologian and paleontologist Pierre Teilhard de Chardin (1881–1955) underscored the special nature of this divine acquisition, which he called "noogenesis."[25] The abstract nature of the mind has led to other philosophical and religious concepts: survival of the soul after the death of the individual as well as its reincarnation in other lives (animal or human). These beliefs have underlain and continue to underlie the ethical stances and spiritual lives of billions of people. To some, they allow either the hope for an eternal, better life after death or the prospect of a reincarnation in another life form. They make some fear negative repercussions for subsequent existences if they lead bad lives in the present; for them, our behavior is guided by our soul and the soul will be judged.

For atheists or agnostics, self-consciousness and knowledge acquired in the course of a life are elements accumulated in our brains, a simple hard drive equipped with a perfect operating system. The soul would be but a by-product of self-consciousness, a simple reassuring concept for easier passage into the realm beyond life, allowing us to live in peace without being endlessly tormented by our unjust, unavoidable end. For these nonbelievers, self-consciousness and the soul (or mind) are unfortunately bound to disappear definitively after the extinction of the physiological current—life—that animates us. This conception of our existence leads hedonists and epicureans, acolytes of Greek philosophers (Aristippus and Epicurus), to take fullest advantage of every day of their existence, which they see as having no tomorrow. *Carpe diem!*

In all these cases, many people aspire to leave permanent imprints of the knowledge they have acquired or of their physical passage here on Earth. In fact, from our lives, from our personal experiences and knowledge, what will subsist will be whatever we have been able to transmit orally or by tangible works that will resist the ravages of time more or less successfully.

The first real evidence of knowledge acquired and transmitted to us by prehistoric man is represented by:

- the first worked stones (about 2.6–2.5 million years ago), for which the workmanship was standardized and continually improved through identifiable stages.
- the first hearths demonstrating mastery of fire (790,000 years ago).
- the first manifestations of social activities associated with death: funeral rites and offerings placed in sepulchers (100,000 years ago).
- the first cave wall paintings (32,000 years ago).

Each of these stages, proven by dated remains in various regions, marks the ascendancy of memory, communication, and self-consciousness. It is estimated that it was just 35,000 years ago that *Homo sapiens* became a self-conscious thinker. He *who knows that he knows* was named *Homo sapiens sapiens*.

So we certainly descend from the Cro-Magnon tribes, prehistoric men whose shape and cranial volume were similar to our own and who left all the types of evidence I have cited. But before them, other, different human tribes from parallel evolutionary lineages (like the Neanderthals) had acquired conceptual techniques for working cut stone, mastering fire, and forming ideas about and rites for death (they left sepulchers). Can we believe that they were already conscious of their identity?

The first rock paintings, dating from 32,000 to 10,000 years ago and produced by the Cro-Magnon (whose origin is estimated at 100,000 years ago), are the oldest evidence of the acquisition of identity and perception of existence. For the first time, a living being deliberately transmitted images of his environment (drawings of dangerous animals or those useful as food), of his hands (imprints), and, more rarely, of himself (hunting scenes and drawings and statues of women). In the many caves where these men displayed their work, we find the same techniques and identical motifs and signatures. Artistic knowledge was transmitted in a vast region where another human tribe survived—one that was less advanced, less educated: the Neanderthals. This tribe disappeared suddenly around 25,000 years ago, while the Cro-Magnon, with their social organization, many tools, beliefs, and morals dominated everywhere for several tens of thousands of years. They therefore hold the longevity record for human organizations, which at that time were characterized by small nomadic tribes of friends or enemies.

The prehistoric paintings in the Chauvet Cave are so well conceived, the effects of superposition of animals and mastery of perspective so technically adroit, that we are forced to recognize the existence of a transmitted art (plate 2). The creators of the frescoes in the Chauvet Cave were above all highly talented artists; certainly they did not display their first work there. Obviously, the mastery of their art was refined during several preceding generations at sites exposed to the vagaries of weather or in caves that have collapsed or are unknown to this day. These men, accumulating and transmitting the first knowledge, had surely acquired self-consciousness!

Henri has allowed us to profit from his reflections on the motivations of these prehistoric paintings. He has enlightened us on the first tangible symptoms of the humanness of our human ancestors. He has made us dream about the distant past of the Cro-Magnon tribes with their artist-shamans evolving in a virgin nature, but a very hostile one, in which man had to survive and reproduce. I have been able to imagine their savage manners, cruel or engaging, suggested by an object or a fresco. Through their works, I have sought to detect their level of self-consciousness.

These reflections make one dream a bit about oneself and one's anonymous neighbors, because these first men, who have left traces of their vision of the world, are our common ancestors.

I have visited many sites where vestiges of prehistoric man are displayed. The South of France is full of them. Each time, I have observed the public fascination with places the Cro-Magnon frequented. In groups of visitors led by guides, silence, whispering, questions, and admiration reflect surprise and emotion. Children stare wide-eyed in the presence of these artistic creations of prehistoric men. We were like that? The dates no longer matter. It is difficult to grasp the differences between tens and hundreds of thousands of years. The gap in time between them and us is so great that it is difficult to believe, to imagine such an ancestry.

I remember a visit to the cave of Pech-Merle in the Lot valley in France where we admired the rock paintings and the moving imprints of children's feet (fig. 1). These little footprints, at the base of a vessel, which possessed an astonishing freshness, struck my son, then 12 years old. He exclaimed, "You see, Cro-Magnon children liked to splash and track mud too!" And a visiting woman hypothesized that, if little children had accompanied the cave painters, these latter could only have been women.

I also recall visits to prehistoric vestiges found at the very heart of my city, Nice, like the grotto of Lazaret, which harbors many remains from human feasts 160,000 years ago, and the site of Terra Amata, where traces of a 400,000-year-old camp have been discovered. I participated in excavations at Terra Amata when I was a student. To exhume these fragile vestiges, I scratched the earth of our ancestors with a screwdriver and toothbrush. It was there, at Terra Amata in my hometown, that one of the most ancient human hearths again saw daylight.[26]

The oldest fossils of human or prehuman remains and their oldest vestiges attest to the slow metamorphosis from animals to man who thinks then expresses himself and communicates. The basis of the natural selection of the Cro-Magnon was the nomadic behavior that led to the isolation of families equipped with better memories, allowing more complex thought and better expression. This evolution is thus linked to the ability of prehumans to move, to leave their initial niche, to conquer other environments, other spaces, to adapt to different conditions of life. It was

Figure 1. A child's footprint in the prehistoric cave of Pech-Merle, France. (Photo from Patrick Cabrol.)

therefore among the boldest, the most courageous, those who braved the unknown, that the dominant lineage emerged.

Beyond Our Ancestors: The Origin of Life

For me, the scientific version of Genesis and the spiritual and symbolic biblical descriptions with Adam and Eve that have so marked the entire Western world for the last 30 centuries are remarkable natural or religious histories transmitted by men who received divine revelations or accumulated earthly knowledge.

For biologists, the scientific discoveries about our recent origin have a logical sequel. Evolutionary mechanisms that allow us to understand the relationships between ape and man also operated between the ape and the smallest independent animal cell. However, these mechanisms do not answer other fundamental questions about the origin of life. How and where did life appear? How did plants and animals arise from bacteria, the first forms of life to appear on Earth? Is there a strong thread conducting life through the ages and hazards from its origin to us? These stages from the birth of life on Earth, sometimes still mysterious, lead to other questions or hypotheses about our future, about what will become of the lives that surround us and about the existence of other forms of life still unknown.

To understand our origins, to foresee our future, to explain the nature of nature—these were also the concerns of the Greek philosophers.

We have to imagine new histories of life because of new, well-established scientific knowledge. These contemporary visions would have been inconceivable in an age of mythic tales and revelations. They are both in accord with known facts and swept of religious, philosophical, and ancient beliefs. But the approach has remained the same since the dawn of time: start with firm knowledge accepted by the great majority of the educated community, then erect hypotheses, speculations, that will be tested by the science of the future. The new conceptions of the history of life allow us to understand the past better and to try to predict the future. They can always make us dream, and they leave a role for the spiritual.

CHAPTER 2

On the Origin of Life on Earth

The First Genesis: The ALH84001 Affair

The mystery of the origin of life on Earth has spawned many hypotheses. As does the study of our first ancestors, this subject also has a mystical and philosophical aura. It entails the origin of the first cells, the initiation of life perpetuated by reproduction, the essence of life of all the species comprising the fauna and flora. It is surely the most exciting subject in biology; it holds surprises that can change our conception of life and orient our vision of our future.

Creating Life in the Laboratory

To explain the appearance of life and to understand the mechanisms of its genesis, several teams of researchers have, since 1924, tried to produce life from inorganic molecules.[1] They have sought the appropriate proportions of the various constituents of living organisms and have simulated the ancestral environment. Heat, electric arcs, an atmosphere lacking oxygen and composed of unbreathable gases, carbon compounds, minerals in aqueous solution, and pressure have all been used in the flasks of these apprentice Creators.

In the past, alchemists attempted in vain to transmute lead into gold in obscure medieval alcoves; today, biochemists, crowned with diplomas and confident in their science (which reduces life functions to molecular interactions), work relentlessly to find the secret of the transmutation of the inorganic into the living. To discover the first stage of life, they use ultramodern laboratories equipped with impressive analytic robots. But the Supreme Master has never deigned to guide them. Their small steps to resolve the enigma are limited to a few ambiguous results that can be summarized in one page.

Since 1953, we have known that without DNA (deoxyribonucleic acid, a molecule shaped like two serpents intertwined in a helix) there would be no life as we conceive it. This very complex molecule is the repository of genetic information and one of the main constituents of chromosomes. It allows life to replicate. DNA therefore constitutes the embryo of life. Now, despite the progress of science and technology, despite perfect knowledge of the structure and composition of the DNA molecule, biochemists have not yet succeeded in recreating the least active gene beginning with a soup of inorganic or gaseous ingredients. At best, they have obtained amino acids, elementary organic molecules that form proteins, normally fabricated according to genetic information.[2] But amino acids and proteins are inert: by themselves, they do not yield life. It has been proven only that the subproducts common to all life can be produced by chemical reactions independently of life.

In the 1960s, other scientists suspected that the complex structure of DNA would be elaborated from a simpler primer (RNA, or ribonucleic acid) that was able to synthesize proteins and reproduce itself. This line of thought suggested that RNA could have been formed by chance, as a spontaneous molecular reaction, and would have the key traits of replicating itself and/or improving itself by natural selection.[3] On the basis of this idea, researchers today are attempting to obtain RNA from inorganic precursors, hoping that the presence of certain minerals will stimulate (catalyze) its formation.[4] In this vein, scientists have observed that this primer could have been created in a rocky or clayey matrix.[5] This last hypothesis recalls the much more mythic one of God creating man in his image with a bit of earth.[6]

To solve the puzzle of the initial prototype of life, the imaginations of other researchers have turned to the matrix of the primitive organism. They have sought to demonstrate that the walls of the most archaic cells were formed as an emulsion. To see this, we have to imagine in coastal lagoons a soup of basic ingredients that could have combined to form organic matter. This process could have begun with carbon-containing molecules present on Earth or coming from outer space. This soup would be concentrated by evaporation, then transformed into a "hot broth."[7] Mixed by wind or fluid currents or gaseous vapors from the ground, this fertile broth would have been conducive to the formation of organic droplets resembling oil in a vinaigrette. A fine film would have isolated the contents of each droplet. The film would be transformed to form a more solid membrane that supposedly would have initiated and protected an unknown chemical process leading to RNA or DNA.[8]

So far, all these speculative scenarios have led to impasses. None of the molecules carrying the information characterizing life and having the ability to reproduce themselves (DNA and RNA) have been constructed in the laboratory from simple organic molecules.

All of these efforts are based on chemical reactions in varied environments as the basis of the creation of life. Christian de Duve, Nobel laureate in biology, has extensively discussed all the highly technical current knowledge about organic chemistry that suggests how cellular life could have begun.[9] However, it is helpful to separate these exciting scientific advances from their place and time; clearly, even if a laboratory fashions a synthetic, functional gene from inorganic molecules, the alchemists of life would never be able to prove that it was formed this way on Earth or elsewhere at a specific time in the past. The mechanisms, still undiscovered, that scientists are seeking concern the universal elaboration of life.

The Oldest Vestiges of Life

Another way to approach the mystery of the origin of life is to seek the initial forms of life that appeared on our planet. To do this, we must examine the oldest terrestrial rocks formed after the cooling of fused substances. Dating from 3.85 to 3.4 billion years ago, they are composed of silica-bearing sediments and carbonates, rarely associated in remains of living organisms. These shrouds of the dawn of life have undergone compression and intense heat that have usually left them obliterated or deformed. The fossils are often blurred and unconvincing, so they are contested by other researchers. As dating techniques for these rocks have been refined over the last 15 years, about 50 scientists continue to compete either to be the first to have discovered the oldest form of life or to win the most fame by arguing about everything that appears on this subject.[10]

The oldest and most surprising of these traces, found on the Greenland coast, date the arrival of life on Earth at 3.85 billion years ago.[11] Measures of the proportions of isotopes of carbon allow researchers to attribute a biological origin to carbon-bearing matter detected at the heart of these rocks. The authors of this discovery, who published their findings in 1996, were astonished to have linked these vestiges of biological origin to a period when the Earth was believed to be very inhospitable. In fact, more than 700 million years were surely necessary for the Earth, a ball of rocks undergoing fusion that was formed 4.56 billion years ago, to have cooled on the surface enough to allow the formation of aqueous habitats suitable for the emergence of life. Moreover, 3.85 billion years ago, the Earth, with no dense atmospheric shield, would have been bombarded with meteorites, each of which could have annihilated nascent life. It is therefore surprising to deduce that life arose in such difficult conditions and left enough traces that researchers, even very astute ones, would have been able to exhume them 3.85 billion years later. This unique observation was strongly challenged in 2002.[12] It rests uncertain and must be confirmed.

The oldest fossilized vestiges of life identifiable under the microscope date to between 3.5 and 3.46 billion years ago. These are traces of multicellular filamentous bacteria or microscopic spherical bacteria found stuck in siliceous rocks (types of chert) from South Africa and Australia (fig. 2).[13] Some of these fossils have also been challenged: could they not simply be traces of infiltration by hydrothermal fluids rich in organic matter or self-assembled structures?[14]

By contrast, the biological origin of rocks more than three billion years old, found in the same parts of Australia and South Africa, are uncontested. These consist of an accumulation of calcareous particles deposited and trapped between filamentous bacteria (related to plant-like bacteria), forming a carpet or sheet (plate 3). Sectioned, these rocks, formed by living organisms, resemble a sliced red cabbage or baklava. This appearance is generated by a discontinuous growth of the bacterial carpet: growth phases alternate with phases in which calcareous particles are trapped. We call these relics of life "underwater petrified gardens" or stromatolites (which means "carpets of stone").[15] Although some very old sedimentary rocks can be mistaken for stromatolites,[16] 48 deposits of true relics of past life dating to more than 2.5 billion years ago have been logged, and the number continues to grow.[17] The oldest are conical protuberances 3.49 to 3.45 billion years old found in Australia at the same site where petrified microfossils have been found in chert (plate 3a).[18] Another well-preserved fossil reef of stromatolites dating from 3.43 billion years ago was found in Australia; extending for more than 10 kilometers, it includes different structures that are strong evidence of its biological origin (plate 3b).[19]

These strange, massive formations are still constructed today by living bacteria residing especially in small soft bottoms of bays in Australia (plate 4a) or under deeper water (as in the Bahamas; plate 4b).[20] They look like dark rocks resembling giant mushrooms growing from beaches of coral

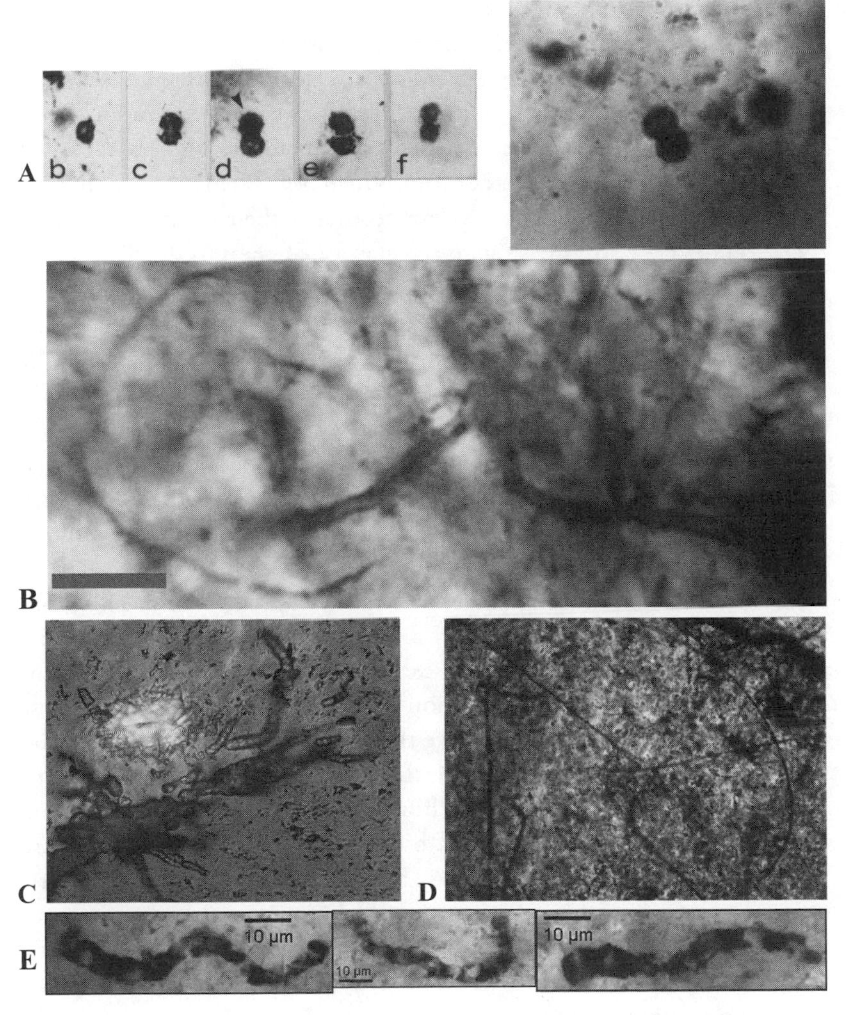

Figure 2. Composite of images of the first microfossils, dating from more than three billion years ago. *A*, Unicellular forms with divisions. (Photo courtesy of Andrew Knoll.) *B*, Filamentous forms containing carbon. (Photo courtesy of Maud Walsh.) *C*, Filamentous forms containing carbon, trapped in vitrified lava. (Photo courtesy of Neil Banerjee.) *D*, Filamentous forms in volcanic deposits in deep sulfurous water. (Photo courtesy of Birger Rasmussen.) *E*, Filamentous forms resembling cyanobacteria, found in chert. (Photos courtesy of J. William Schopf.)

sand. These contemporary stromatolites illustrate schoolbooks as realistic depictions of the oldest evidence of life on Earth. The bacteria that form these rocks today are mainly represented by several filamentous species (composed of several cells in a line),[21] and they have the same mode of nutrition as plants (photosynthesis): water, carbon dioxide, mineral salts, and energy from sunlight act together to produce organic matter and, finally, oxygen as a by-product. Their special feature is their ability to produce rocks. Today, we still investigate the nature of the rare fossilized bacteria found in the oldest stromatolites (bacteria functioning by virtue of chemical reactions with minerals or, more likely, like today's plant-like bacteria, by photosynthesis). One fact is agreed upon by the scientific community: if the bacteria building the first stromatolites were photoautotrophs (able to grow by photosynthesis), when they became common the free oxygen gas they generated could have begun to oxidize dissolved iron and sulphur, producing precipitates of rust (Fe_2O_3) and sulfate (SO_4), respectively. Scientists have thus sought the age of the first oxidized rocks. The largest deposits of oxidized sediments date from between 2.5 and 2.3 billion years ago and are found in many strata around the world. This global distribution has been interpreted as the consequence of strong initial photosynthetic activity on a global scale. But the first (rare) oxidized minerals have been dated from 3.8 billion years ago, and the nature of these first traces of oxygen is debated—are they biotic or abiotic? Authors who support the hypothesis that biological production of oxygen is very ancient believe that for more than a billion years (between 3.5 and 2.5 billion years ago), most minerals oxidized by the first plant-like bacteria were reduced by the hot, acid water ejected by many volcanoes over large parts of the Earth.[22] As part of this argument, the question also arises about whether the first plant-like bacteria could have been very localized for a long time, spreading very slowly and therefore not leaving widespread traces of oxidation.

Some paleontologists believe that the remarkable stability of the massive structures of stromatolites through the ages was accompanied by an even more surprising stability of the organisms that constituted them. A metaphor has been used to describe this double absence of evolution: it is not the "Volkswagen syndrome." In fact, the remarkable persistence of the chassis of the Volkswagen beetle was accompanied by substantial technological progress on the motor through the years. By contrast, the absence of evolution of the carpet of stones did not mask an internal evolution of the organisms that constructed it; they are still multicellular bacteria (the most common plant-like bacteria today are cyanobacteria).[23]

Such a precocious appearance of stromatolites, composed primarily of highly structured filamentous bacteria (many cells stacked on one another), is a crucial element in understanding the appearance of life on Earth.

Other traces of filamentous bacteria, from 3.235 billion years ago, have been discovered in Australia (fig. 2d).[24] They had a different mode of life, developing around sulfurous hydrothermal springs. Therefore, they do not need light and can live at great depths. Similarly, traces of tubular microorganisms approximately 0.2 millimeters long, dating from 3.5 billion years ago, have been found in volcanic lava from South Africa (fig. 2c).[25] Finally, microbial sulfate reduction was established at least 3.47 billion years ago.[26]

All descriptions of traces of life from 3.5 to 3 billion years ago are from the past 20 years, and some of them must still be confirmed. But the most respected paleontologists, who sometimes dispute the nature of certain early traces, agree that life existed on Earth at this time. Therefore, even if some pieces of evidence bearing on early, ancestral life are controversial, it is clear that the mass of evidence currently available converges to produce a strong argument for the presence of life at that early date. Now that the beginning of life on Earth by 3.5 billion years ago has been established, we can formulate two observations.

The first concerns the estimation of the real time when life appeared on Earth. Paleontologists rely on the discovery of chemical traces, remains produced by bacteria, imprints of fossilized structures. It is dating these features that fixes the oldest limit at 3.5 billion years ago. But this conclusion must be tempered when we take two facts into consideration. One the one hand, deposits of rocks dated from 3.8 to 3.4 billion years ago are extremely rare, and their geographic distribution is highly restricted. The great majority of the surface of the Earth of that period has been reworked by movements of the crust. Rocks and sediments of that epoch have therefore been uplifted, buried, compressed, heated, and melted. On the other hand, the formations that are the richest in vestiges of life and that date from 3.5 to 3.4 billion years ago are coastal regions or springs of hot, sulfurous water. Now, the coastline and these springs represented at that time just a tiny oasis of life in comparison to the enormity of the oceans and emergent land. These facts suggest that the very rare deposits with traces of life dating to 3.5 to 3.4 billion years ago that have fortuitously been discovered so far surely do not represent the first zones colonized by life on Earth. The first colonization of the planet at one or a few extremely localized points must have occurred much earlier, and we cannot estimate with any precision the age of the initial colonization with only the vestiges found to date. Thus we can only say today that life appeared on Earth *at least* 3.5 billion years ago.

The second observation concerns the ensemble of traces of life from 3.5 to 3.4 billion years ago. This is the strange fact that the ancestral bacteria were already highly diversified at this time (unicellular, filamentous, multicellular, forming dense carpets and structures on the coast or growing near sulfur springs). The presence of all these forms between 3.5 and 3.4 billion years ago argues for an older origin with simpler, less diverse forms.

In sum, all these vestiges of very ancient life provide no precise clues to the place, time, and mechanism of the genesis of the first living organisms. Therefore, the currently popular idea that life probably arose in warm, subsurface waters along a mid-ocean ridge, the kind of environment where a great variety of heat-resistant bacteria thrive today, is a hypothesis without any scientific basis. We know only that between 3.5 and 3.2 billion years ago the diversity of bacteria was established in varied conditions. Saying that one kind of bacterium (sulphate-reducing or photoautotroph) or one condition on Earth (near deep, hot springs, or on the subsurface) was at the origin of life is speculation.

The Smallest Known Life Form

Another element of the problem must be considered. Since bacterial life appeared on Earth, no intermediate, more archaic form—fossil or living—has been found. We have to eliminate from this category the viruses that are parasites of bacteria or animal or plant cells, unable to live without them because they are unable to grow (as are all parasites) and unable to replicate (a specific inability of viruses that use host DNA). As is true with most parasites, viruses must be considerably simplified relative to their ancestors; their structure and means of growth and reproduction have become highly adapted to their host, to the detriment of the latter. Viruses cannot live alone, so their current form cannot be considered the model for the precursors of life. However, viruses may be atrophied descendents of bacteria, or even of unknown forms of life simpler than bacteria.[27]

Bacteria are therefore the simplest living organisms known today. Since the appearance of bacteria on Earth, certain evolutionary pathways have allowed complex, structured forms of life (fauna and flora) to emerge. This process has never eliminated the simple forms like bacteria, so numerous on Earth today and so well adapted to available environments. All evidence points to the fact that the simpler life is, the better its chances of adapting to various environments and persisting. There should therefore still exist today an abundance of reproductive organisms intermediate between prebiotic matter and bacteria. New bacteria should appear nowadays by evolution of these simpler forms. But no transition process between inorganic matter and bacteria has been found in nature. No prebacterial organism able to survive independently (without using a more structured organism) has been identified among species, either living nowadays or fossilized.

Of course, potential leads have been studied, like the mysterious nanobacteria, recently discovered in great numbers on rocks and living organisms (there are even some in our blood). These are spherical or oval entities that can be observed under the electron microscope. Their name refers to bacteria, but only by analogy, because their external shape *evokes* bacteria. Their volume is 100 times less than that of ordinary bacteria. Their size is in the range of several dozen nanometers (a nanometer is equal to 0.000001 millimeter), thus, their prefix. We do not yet have a good idea of what they contain and how they multiply. Are they simple crystals, fragments of mineralized bacteria, or the first germs of life? In fact, we do not know much about the origin or function of these particles, and the few biological hypotheses about them are controversial.[28] Finally, there are no intermediates between these nanobacteria and bacteria.

While we wait for elucidation of this line of research on the smallest living organisms, the problem of the absence of transitional stages between undoubted prebacterial structures and what are clearly bacteria must be considered. For this reason, an international panel of specialists was convened in 1998 under the auspices of the National Academy of Sciences of the United States to consider hypotheses about the lower size limit of microorganisms. This panel concluded that the smallest diameter of living organisms is between 0.0002 and 0.0003 millimeters (equal to 200 to 300 nanometers or 0.2 to 0.3 microns). The length of the smallest bacteria known today is about 0.2 microns.[29]

Other scientists have sought the minimum number of genes required by a functioning organism able to reproduce itself. For example, based on the known bacteria that possess the least DNA, approximations suggest the minimum number of genes required to code for vital proteins is about 200.[30]

But how has life advanced from nothing to the genetic capital contained in an organism of 0.2 to 0.3 microns?

This question led to the idea of a unique, universal ancestor of all bacteria, named LUCA (last universal common ancestor).[31] This view of the origin of bacteria stipulates that there was only a single exceptional event—some authors speak of a miracle—giving rise to a single bacterium that has subsequently reproduced and diversified by evolution. In fact, it is more likely that the soup of complex organic molecules produced several types of bacteria, of which only certain ones led to the origin of the bacteria that appeared on Earth more than 3.5 billion years ago. That is, instead of the catchy image of a single bacterium as the origin of all life, it is above all the processes occurring at the time that should be emphasized.[32] Most likely there was a swarming of bacterial life formed from the association and union of organic molecules. This process of the genesis of bacteria resembles the workings of an automobile factory: all pieces necessary for the assembly must be in place and the mainspring of the system (RNA and DNA) must be able to replicate itself and the intermediate assemblages, as well as the final assembly. To limit the output of this extremely complex chemical system to a single product (LUCA) is to ignore the immense potential of the processes operating during the appearance of the first bacteria, which inevitably differed from but greatly resembled each other.

It is a fact that, at the beginning of the third millennium, we cannot yet describe and illustrate the processes and the stages in the genesis of bacteria. The exact time and place of the spontaneous generation of the first bacteria also remain unknown. We must concede that bacteria—prodigiously complex biochemical factories, endowed with a system of self-replication, containing DNA and RNA, themselves very complex molecules—have no identified ancestors. Bacteria remain both the smallest living structure known and a marvel of biological and chemical technology. We can view them as the original seeds of life on Earth.

There are but two possibilities to explain the lack of evidence on the transition between inorganic matter and bacteria, archaic forms of life that are nevertheless complex.

1. The intermediate forms of life or the environmental conditions favorable for the process producing bacteria have disappeared completely on Earth without leaving any traces, for unknown reasons. We can imagine, for instance, that a catastrophe (like a meteorite impact or a volcanic eruption) struck the restricted zone in which they lived. Only the bacteria survived. They must have been more widely distributed around the globe or, more likely, better equipped to survive the catastrophes or other quick environmental changes.

or

2. The transitional forms between inorganic matter and bacteria have never existed on Earth. Bacteria, well-adapted to survive the conditions of cosmic voyages, arrived alone on our planet, natural selection having eliminated the intermediate, less resistant forms during the voyage.

Astrobiology and the Dogma That Life Originated on Earth

We must humbly recognize that the first hypothesis, that of the birth of life on Earth, is only an unsupported hypothesis; all research trying to confirm it is at an impasse. It is just an idea that seems logical and evident. An idea that is taught. This idea has become a dogma.

These considerations lead us to examine the alternative hypothesis more closely, that the seeds of life that appeared on Earth came from elsewhere. *Elsewhere!* This is a completely different view of the matter that implies the existence of incubators of life elsewhere and other unknown forces behind the origin of the first gene and the first process assembling molecules in a way that led to bacteria. This is an easy solution—the problem of the origin (the place and time of the genesis of life) is unresolved; it is displaced. But the idea is seductive; it is also based on nothing, it is just another speculative hypothesis. It has become a theory. Two Nobel laureates are among the pioneers of this theory: the Irishman Lord Kelvin (1824–1907) and the Swede Svante Arrhenius (1859–1927), who gave this theory the name *panspermia* (which literally means "seeds disseminated everywhere").[33] Those who work on this hypothesis are exobiologists (biologists who seek life elsewhere) or astrobiologists. It is an iconoclastic idea, just as disturbing as the ideas of Nicolaus Copernicus, who by 1513 had affirmed that the Earth revolved around the sun (reaffirmed in 1615 by Galileo, with no more success in his attempts to communicate it), or Lamarck's idea in 1800 of an ape transformed into man (reaffirmed in 1871 by Darwin, who succeeded in convincing a circle of scientific disciples).

The theory of panspermia was defended by many Nobel laureates during the second half the twentieth century. The most imaginative among them was Francis Crick (1916–2004), codiscoverer of the structure of DNA. Crick wrote a book in which he propounded his hypothesis very seriously; he called it "directed panspermia."[34] According to him, a spaceship arrived from elsewhere, intentionally loaded with bacteria, and crashed on Earth, thus liberating a great diversity of these microorganisms, of which certain types have survived and proliferated. This book is symptomatic.

Crick, discoverer of and specialist in DNA, rejected the hypothesis of the spontaneous formation of DNA on Earth from nonliving matter. He concluded that DNA came from elsewhere. Not yet aware of the "contaminating" ability of meteorites, he imagined the inoculation of Earth by an extraterrestrial power, going so far as to describe the characteristics of the spaceship bringing us life. Works of science fiction could have inspired this hypothesis! Applying the logic of the conquest of space, which envisioned sending bacteria to Mars in order to create a breathable atmosphere, we can imagine that an extraterrestrial civilization fertilized Earth to render it hospitable by creating, through the proliferation and evolution of bacteria, an oxygenated atmosphere (thanks to photosynthesis) and a suitable environment. Will they land here one day to gather or capitalize on the fruits of their sowing?

A Meteorite from Mars?

In 1980, Imre Friedmann (1921–2007) was a professor of biology at Florida State University in Tallahassee. I had invited him to Nice to participate in the jury assessing my doctoral thesis. He was then a specialist in Caulerpales, an order of strange green algae to which I had devoted 10 years of study. He subsequently specialized in microalgae, then in bacteria of cold environments. He became the leading specialist of the microorganisms of the Antarctic ice.

At this time in the United States NASA (the National Aeronautics and Space Administration) was developing an interest in the planet Mars, with two recurring questions: Is there life there? Did life formerly exist there?

However, Mars is a cold planet with a thin atmosphere composed mostly of carbon dioxide, where water was abundant in the distant past and is still present as surface ice. Now, water was the medium in which life arose on our planet. It is therefore logical to seek evidence of a parallel or similar development of life on Mars. While NASA succeeded in placing robots on Mars to gather data on the physical and chemical characteristics of the soil and the atmosphere of the red planet, a meteorite found in 1984 on the Antarctic ice floe aroused more and more interest. It is a modest rock, weighing barely 1.9 kilograms, but appearing very different from other extraterrestrial rocks (fig. 3a). It was cut up into fragments that were sent to various specialists. Friedmann, one of the few experts on Antarctic bacteria, was asked by NASA to detect eventual contamination of the meteorite by bacteria of local origin—that is, from Earth.

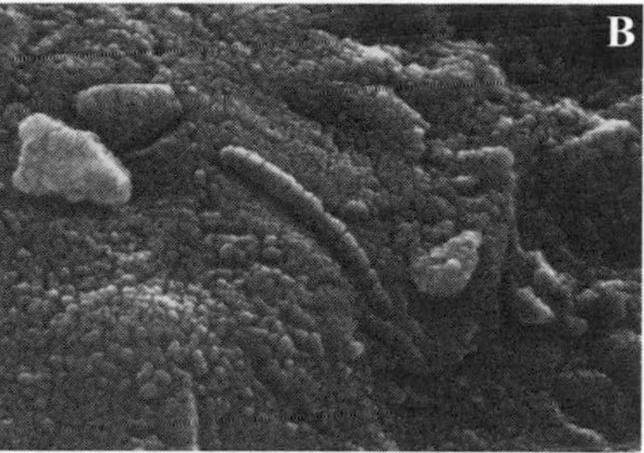

Figure 3. *A*, The meteorite ALH84001. *B*, The strange vermicular forms observed on the meteorite ALH84001 resemble common bacteria but are still unexplained. Claims that they strongly suggest primitive life from Mars that were made in 1996 in the journal *Science* have been contested. (Photos courtesy of David S. McKay.)

I have often been impressed by fortuitous encounters that have enriched my life. An unusual sequence of circumstances led to my seeing Friedmann again.[35] Passing through Nice in February 2001, he came to see me, 20 years after having participated in the jury for my doctoral thesis, bringing to my mind all that happened then and all that has followed . . . while I was writing this book. I invited him to dinner at my house. The discussion naturally turned toward his research, and he told me the story of the meteorite and of his remarkable discovery. I recount it here in the manner of a police investigation bearing a code name: ALH84001.

The ALH84001 Affair

Friedmann began by informing me that the rock ALH84001 had been found in a region that he knew well: the Allan Hills (thus, ALH) a range located at the edge of Antarctica. It was the first meteorite discovered in this zone in 1984 (hence 84 and 001). He told me it was the twelfth meteorite believed to have come from Mars—it was found on Earth as the new millennium was about to begin.

I stopped him, pointing out that this tale was beginning badly; already the "believed to" connoted uncertainty. He explained to me that, in fact, there is very little uncertainty. The composition of the meteorite was analyzed by chemists and physicists. It consisted of rock and gas, because meteorites often contain fragments of vitrified rock with tiny gas bubbles. Several techniques are used to date rocks and gas. One measures the percentage of certain unstable radioactive atoms that gradually transform into more stable atoms.[36] In addition, for every fundamental chemical element (atom), one measures the percentage of each of its varieties (isotopes).[37] The percentages of the various stable and unstable isotopes correspond to a "genetic imprint." The characteristics of meteorites have been compared to those of the main elements of Martian soil and atmosphere, gathered by the Viking probes that landed on Mars in 1976, and the atomic and isotopic composition of ALH84001 is identical to that of Mars. These arguments are cogent. I am convinced of the Martian origin of this rock. Moreover, geochemists have not yet rebutted this contention.

The chemical signature also allows one to date the birth of the rock that constitutes the meteorite (geophysicists do this by measuring the natural disintegration rate of isotopes). Surprise: scientists date ALH84001 at 4.5 billion years old, while the planets that revolve around our sun were formed around 4.56 billion years ago. How can one not be as impressed by the geochemists' deductions as by the remarkable nature of the meteorite itself? It is the oldest rock from the solar system that has yet been found.[38]

Friedmann then described to me how this rock was fissured on Mars 4 billion years ago. In the fissures of the meteorite, carbon-based matter was deposited 3.9 billion years ago. These muddy deposits solidified, their contents became fossilized. Again the chemical analysis of the meteorite constituents confirms this.

I am beginning to get used to the exploits of geochemists!

Friedmann then asserted that 16 million years ago an asteroid must have struck Mars, ejecting into space many Martian rock fragments, of which ALH84001 was one.

There you go too far! How can you say that?

Simple: first, it is the only plausible explanation for the ejection of the Martian rock into space. Then, the time spent wandering in the astral void left measurable traces. In fact, the rock was exposed to cosmic rays that modified the atoms on the outside of the meteorite in certain ways. The extent of these modifications allows an estimate of the duration of the exposure to radiation and, thus, of its sojourn in space.

Once again, I am left dreaming of the technical accomplishments of geochemists.

Friedmann continued by revealing to me that the meteorite fell to Earth just 13,000 years ago, landing on the ice of the Allan Hills.

The explanation is simple. It suffices to analyze the isotopes of the gas bubbles trapped in the ice located around the meteorite, and one can determine the year it landed, within a few centuries.

Friedmann explained to me that such meteorites could have fallen to Earth at any time. And at the period when the first seeds of life seem to have appeared on Earth (between 3.85 and 3.5 billion years ago), the environment was very different from today. At that time, there was not the dense atmosphere that today burns up the majority of asteroids before they reach the Earth (shooting stars are just small meteorites that burst into flame and consume themselves when they penetrate the upper atmospheric layers). The Earth was bombarded by asteroids and comets, and it was also pouring out lava, dust, and inert gases. Its surface layers were unstable—it swelled, parts were uplifted. Crusts that formed were swallowed up in immense, deep geological faults only to reappear, transformed by great pressure and heat. All these movements formed great humps (mountains) and depressions in which water accumulated (oceans), erasing the meteorite craters. On the moon, and on Mars, the soil quickly cooled and was definitively fixed; traces of asteroid impacts (the lunar craters that we can distinguish with a good pair of binoculars) were not erased by movements of the ground as they had been on Earth.

It is in this context, in water that was then oxygen-poor, that life appeared on Earth. Birth—creation? Or birth from a model that came from elsewhere, emitted from the heart of an asteroid that had just splashed into a soup of ingredients favorable to life?

To answer these questions, astrobiologists have sought traces of extraterrestrial life in the carboniferous sediments of meteorite ALH84001. Several surprising elements have been discovered in this quest.[39]

Organic molecules (complex hydrocarbons) have been found, molecules very different from those that might have formed autonomously from inorganic matter. Thousands of these molecules are found today on Earth, but not one resembles those analyzed from ALH84001.

Under the electron microscope, examination of the carboniferous part of this meteorite shows elongate, worm-shaped corpuscles 0.0002 millimeters long broken into globules 0.00002 millimeters in diameter (fig. 4b). Bizarre! These forms resemble colonies of present-day bacteria. Scientists who have examined ALH84001 have also noted the presence of very many tiny, isolated, spherical globules measuring 0.00002 millimeters in diameter. These aligned globules and isolated particles have a volume 200 times smaller than modern-day bacteria. They have been classed with the nanobacteria currently found on certain rocks or organisms. But, as I have said, these nanobacteria remain enigmatic; we do not yet know what they are. Simple crystals? Bacterial fragments? Mini-bacteria?

It is, however, on these elements that NASA based its attempt in August 1996 to affirm that life had existed on ALH84001 and, therefore, on Mars.[40] Occurring amidst a period of budgetary consideration (during

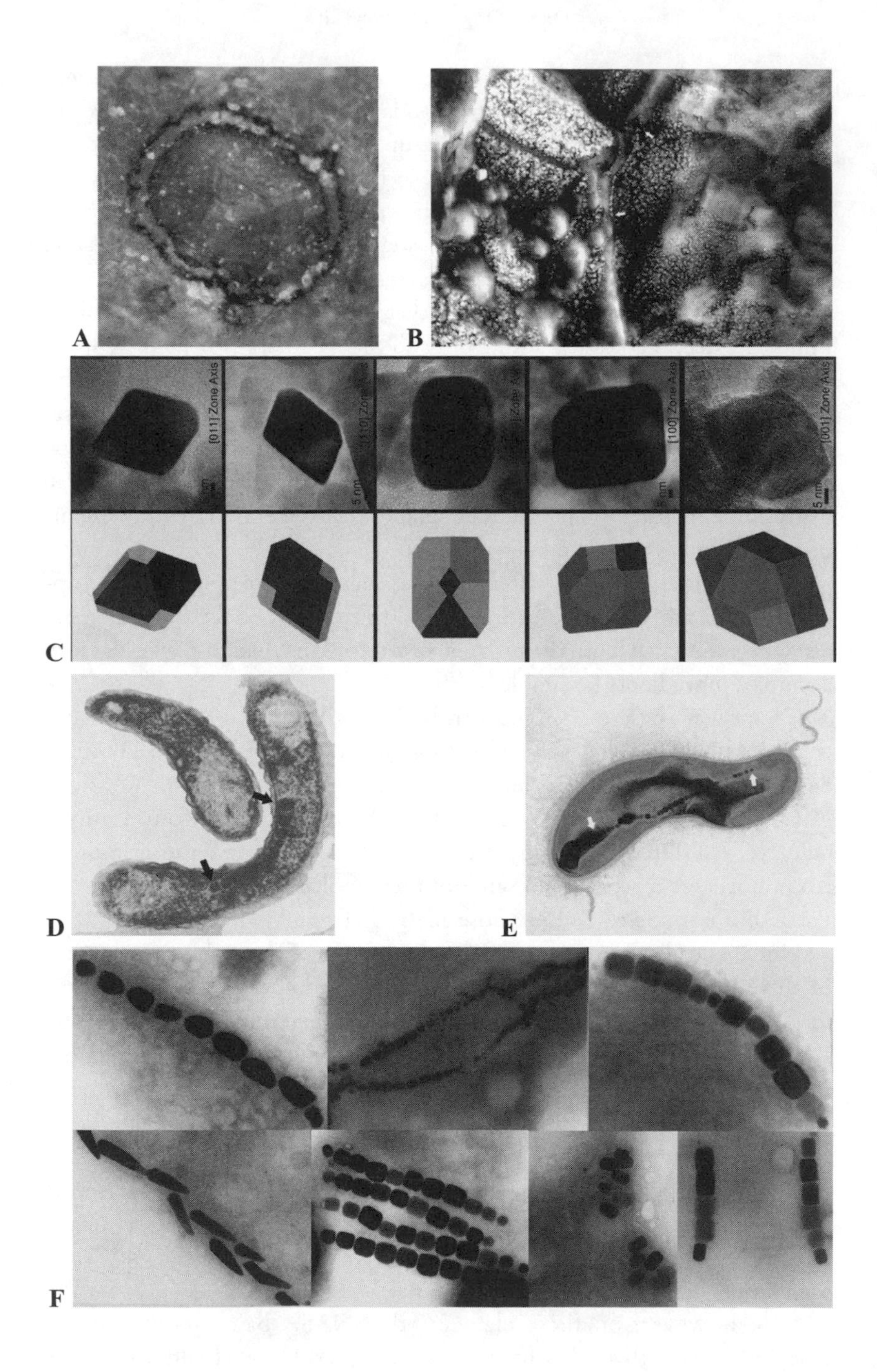
A
B
[011] Zone Axis
[110] Zone Axis
5 nm
[100] Zone Axis
[001] Zone Axis
5 nm
C
D
E
F

which NASA was seeking to increase missions to explore the surface of Mars), this news seemed too opportune and was immediately suspect. Subsequently, other experts strongly contested the elements at the base of this hypothesis, and researchers who had been criticized reacted. It was shown that the aligned globules *could also have been* remains of terrestrial bacteria infiltrated into the meteorite and recently fossilized.[41] The microscopic globules *might not be* nanobacteria, but rather simple inorganic crystals.[42] Research teams have even succeeded in proving that the carboniferous organic matter found in ALH84001 *could have been formed* by chemical means in the absence of living organisms.[43] Finally, contamination by Antarctic bacteria was also suggested.[44]

This crossfire of very technical, argumentative studies, disputing the various elements of the case advanced by NASA in 1996, cast doubt on the NASA hypothesis. It was a fiasco, an immense disappointment for authors, who had been enthusiastic, and for the editors of *Science*, who had been too willing to publish the article. This was a bad setback for NASA as well, and for Bill Clinton, who had had the privilege of announcing the news to the world.

A Surprising Trace of Life

It was at this point that Friedmann had become personally involved. He was tasked with examining the highly publicized Martian rock. Interest in this particular meteorite, far from the only one from Mars, resided partly in its composition (carboniferous zones) and partly in the fact that the internal part of the rock remained sheltered from all contamination. Since landing on the Antarctic ice, an environment very hostile to life that Friedmann knew well, the rock had remained frozen. He therefore focused his attention on certain internal parts of the meteorite: strange

←

Figure 4. Magnetites. *A*, Carbonate globules approximately 0.2 millimeters in diameter from ALH84001 in which hydrocarbons and magnetite chains have been found. (Photo courtesy of Kathie L. Thomas-Keprta.) *B*, Chain of crystals (magnetites) in ALH84001. The configuration of the chain and the highly regular spacing between the crystals constitutes strong evidence for a biological origin. (Photo courtesy of E. Imre Friedmann.) *C*, Magnetite crystals from ALH84001. The regularity of the shapes (truncated hexa-octahedrons) and their size constitute strong evidence for their biological origin. (Photo courtesy of Kathie L. Thomas-Keprta.) *D*, Present-day bacteria containing magnetite chains (between the two arrows). (Photo courtesy of David Balkwill.) *E*, Magnetite chain in a present-day bacterium (between the two arrows). (Photo courtesy of Hojatollah Vali.) *F*, Chains of magnetite crystals (magnetosomes) in present-day bacteria showing very different and specific sizes and crystallization patterns in different species. (Photo courtesy of Dirk Schüler.)

carboniferous globules (fig. 4a). Previous experts had been astounded to find within these carboniferous globules many small crystals of iron oxide known as magnetites—minuscule magnets.[45] Friedmann knew that these magnetites existed in profusion in certain present-day bacteria; they are also found in some cells of birds and in certain human cells (figs. 4d–f).[46] They play a role in our orientation faculty (in learning that, I came to understand why certain orientation abilities are easily lost and other less so: individuals are endowed with these specialized cells, which are more or less likely to be lost through time). Magnetites found in living cells today are always arranged in a line, with each element separated from its neighbors by a fine membrane. When we get a good preparation for observation under the electron microscope, the aligned magnetites resemble an unclasped pearl necklace. Like a necklace posed on a table, the magnetite chain has gentle curves. Earlier experts had noted the magnetites present in their preparations of ALH84001 fragments, but they were scattered or piled up.[47] Such configurations did not suggest they had been formed by a living organism. But the early reports had noted the striking purity of the crystals and their regular form; these recurring traits in magnetites of biological origin are hard to explain as resulting from mineralization (by impact or heating). On this point, the authors who first discovered magnetites in ALH84001 were clear: if subsequent research showed chains of magnetites in the meteorite, this would constitute proof of life on Mars.

This is the subject Friedmann worked on. He suspected a defect in the mode of preparing the fine slices of the meteorite and felt that the procedure used by his colleagues might have dissolved the rock too much. If this was the case, the micro-magnets liberated from their straitjacket would immediately have dispersed or regrouped.

Using other methods, he confirmed his suspicions. In fact, each pile of magnetites corresponded to a beautiful chain of crystals. They were arranged in orderly fashion as in our cells, lined up and separated from each other (fig. 4b). But no inorganic phenomenon is able to produce magnetite necklaces so regularly arranged. Friedmann and his team found that the carboniferous globules of ALH84001 contained many of them. This was not due to chance or contamination; they were contained in the Martian rock when it fell to earth and was isolated for 13,000 years in an Antarctic ice field. Modern bacteria containing magnetites were not able to penetrate the dense, rocky heart of ALH84001.

To prove this, high-quality photographs were needed. But the tiny magnetized crystals confound the signals of normal electron microscopes.

Even with the best preparations, Friedmann obtained only fuzzy, low-resolution pictures. It then occurred to him to find electron microscopy experts. In Germany, some of these machines are made by a famous manufacturer of optical instruments: Zeiss. He obtained the cooperation of the leading engineer in charge of perfecting electron microscopes. For two days, Friedmann was given access to a highly sophisticated experimental model. He took dozens of series of snapshots, one of which was excellent: the magnetite necklaces were magnificent. Finally he was ready to publish his discovery. He was able to show "skeletons" of bacteria embedded in the rock of ALH84001.

Alas, the context of this discovery was unfortunate. Astrobiologists, who had recently been challenged on their first analysis of ALH84001, were increasingly considered only marginally reputable. They were suddenly classed with UFO enthusiasts—people who claimed to have seen (or hoped to see) flying saucers and extraterrestrials land on Earth.

Friedmann and his collaborators presented a preliminary note and a poster at a scientific congress, attracting more incredulity than attention.[48] A more elaborate version of the discovery was sent to the most prestigious scientific journals, *Nature* and *Science*. Burned by the earlier disappointments about the same meteorite, the editors rejected the article without having it reviewed, because the subject was too controversial. Friedmann was despondent and outraged by this unjustified refusal.

At this point, I was reminded of Galileo confronting the ecclesiastical authorities who then ruled science. The motivations were identical: to avoid controversy. It is forbidden to cast doubt on the order of things.

Friedmann then thought about submitting the article to the respected National Academy of Sciences of his country. First, he sent his manuscript to several experts and antiestablishment scientists who had taken up their pens to shoot down the earlier astrobiological articles on ALH84001. In sending them the article, he asked for their critical judgment and the right to cite them in the acknowledgments published at the end of an article. All found his reasoning to be flawless. The article, citing these experts and thanking them for their critical support, was published in January 2001 in the *Proceedings of the National Academy of Sciences of the United States of America*. This journal, though prestigious, is not as well publicized as either *Science* or *Nature*.[49] In the same issue of this journal, another team reported another peculiarity of the magnetites of ALH84001.[50] These authors had focused their research on the nature of the crystallization of the magnetites; according to them, it was very specific, identical in all points to that produced by terrestrial bacteria (figs. 4e and 4f). The authors of this second team arrived at the same conclusion as that of Friedmann's team: *these magnetites should be considered a signature of life!*

Friedmann and his collaborators were then able to savor their discovery, well publicized and received at last. These necklaces of magnetites, veritable fossilized "skeletons" of bacteria, exist in Martian rocks.

One month after publishing this article, Friedmann told me about the difficulties, suspicions, and jealousies he had encountered after publishing his discovery. However, I was speaking to an 80-year-old man, long retired. His status as professor emeritus and his great knowledge of Antarctic bacteria had led him into this scientific adventure. Frail, even timid, sincere, without a laboratory to support or a career to defend, and without public or private benefactors to solicit, he was as surprised and excited by what he had just discovered as he was shocked by the animosity aroused by his discovery.

As I became increasingly interested in the magnetite necklaces, I learned that another meteorite, found in France near Montauban in the village of Orgueil, had also raised many questions as it was studied. This rock, found in 1864, is currently on display in the Natural History Museum of Paris. In 1967, chains of opaque crystals had been photographed in the stretched out corpuscles at the heart of this meteorite.[51] It was not until 1998 that a team made the connection[52] between these and bacterial magnetite chains:[53] the same deductions as for ALH84001! But, in contrast to the rock Friedmann had examined, this one had few magnetite chains. Furthermore, the meteorite had fallen on terra firma and had been exposed to local bacterial contamination to a far greater extent than ALH84001 experienced in the Antarctic ice floes. Friedmann told me that he had also examined a fragment of this meteorite to conclude definitively that the supposed traces of life had not been exogenous.[54]

However, globules similar to the enigmatic nanobacteria and composed of carbonates and cylindrical forms the size of bacteria have been found inside four other Martian meteorites.[55]

Even though some of these observations have been challenged (especially after the evidence of contamination by terrestrial organisms of another Martian meteorite that had fallen in Tunisia[56]), the existence of the magnetite necklaces of ALH84001, fossilized skeletons of Martian bacteria, a presumed signature of life, leads us to recognize other, less well-established observations. What is certain is that the recent discoveries about the Martian meteorites sharpen the curiosity and imagination of researchers. They also force the hand of others who, prudent or overly reserved, have not clearly communicated—or have not communicated at all— observations in accord with the hypothesis of life arriving from Mars.

Life . . . Elsewhere

It is true that minuscule magnets aligned in a thin slice of a Martian meteorite can be disconcerting. They suggest that bacteria, living beings, existed very early on Mars. These traces of life are older than those so far found on our Earth. Furthermore, all modern bacteria that contain magnetite chains live in an atmosphere with at least some oxygen. Today, some such bacteria are even dominant in well-aerated muddy sediments.[57] Therefore, for the bacteria with magnetite necklaces to have survived on Mars, there must have been oxygen. This oxygen could only have been produced in sufficient amounts by chemical reactions triggered by other bacteria containing a highly complex molecule, chlorophyll, which gives a green color to organisms that possess it. Chlorophyll-bearing bacteria currently exist on Earth in great profusion. It is this chlorophyll that allows green plants to live and to liberate the oxygen that we need to live. We can deduce, therefore, that there were probably bacteria with chlorophyll on Mars more than 3.9 billion years ago.

If we consider all these facts, we can imagine the appearance of life on Earth through the arrival of a collection of bacteria, already highly evolved, from Mars or elsewhere in a meteorite similar to ALH84001.

This chain of reasoning can be shattered by one commonsensical remark, however: the bacteria from elsewhere had to have survived for some time in the cosmic void before having sowed the seeds of life on Earth. Is such a thing even possible?

We have long known that certain bacteria, with or without chlorophyll, can become encysted. These cells can "hibernate" or "sleep" for several years. Beforehand, they surround themselves with a thick carapace able to resist harsh environmental conditions. More recently, we have learned that some bacteria are able to "sleep" in this fashion for tens of millions of years. In 1995, abdominal bacteria from a bee fossilized in amber dating from 25 to 40 million years ago were able to be "resuscitated."[58] In October 2000, other researchers drew samples of crystalline salt from 600 meters below the Earth's surface. These crystals, formed 250 million years ago, contained small liquid lenses in which bacteria had been trapped. These bacteria had survived and began growing and dividing as soon as they were freed from their salt sarcophagus.[59] These recent discoveries of "resuscitated" bacteria have been contested,[60] but some of the original authors remain convinced of their position.[61] The fact is, respected bacteriologists believe it is possible, and prestigious scientific journals using highly qualified referees have published these remarkable discoveries.

Another discovery illustrates the extreme hardiness of some bacteria. It concerns a bacterium, *Deinococcus radiodurans*, that was discovered in 1956 and grows in a vacuum, survives dehydration, thrives in the vicinity of intense radioactive sources, and even manages to reconstruct its genome after it has been broken into a hundred fragments.[62] Craig Venter, director of the private laboratory that published one of the first studies on the complete human genome sequence, participated in the sequencing of the genome of *D. radiodurans* and considers this bacterium a possible example of interstellar tourism.[63]

Finally, we must also ask about the landing conditions. Can bacteria survive the great heat during entry into the atmosphere and the shock of hitting the ground? To answer these questions, researchers have been able to prove that the carboniferous substances contained in the meteorite ALH84001 were not exposed to temperatures above 400°C. There was therefore no sterilization upon landing.[64] Also, another team undertook ballistic tests with rocks hurled at speeds of 300 to 600 meters per second and concluded that meteorites would clearly be able to land without exploding or being consumed before reaching earth.[65]

Thus, we can now admit that certain very resistant bacteria, endowed with a casing as airtight as an astronaut's suit, protected from cosmic rays in the deep interstices of an extraterrestrial rock, would in fact have been able to wander safely in the astral void, fall to Earth without being burned up, and land alive on our planet. Those that happened to land in propitious conditions, such as an ocean or swamp rich in carboniferous matter (which is chemically similar to organic matter) and minerals, would indeed have been able to grow.

Moving from speculation to still further speculation, we can also ask about the origin of life on Mars. Though we accept the reasoning leading to the estimate that bacterial life flourished on Mars 3.9 billion years ago, it is difficult to admit that this complex form of life evolved on that planet in just a few hundred million years. We arrive at this figure by considering the time when bacterial life might have arisen (as on Earth, the surface rocks of Mars refroze 4.2 billion years ago) and the time when it was already flourishing (more than 3.9 billion years ago, if we recall the bacterial presence in Martian meteorites from that time). We can thus imagine the appearance of life on Earth by the arrival of a group of bacteria that were already highly evolved, from Mars or *elsewhere,* but that had been born and had diversified *elsewhere.* A life created elsewhere, coming from another solar system, associated with another star that appeared well before our sun! The big bang, at the origin of our universe, dates to 13.7 billion years ago. Our universe is now composed of billions of galaxies, themselves each composed of billions of stars surrounded by planets. This constitutes a huge amount of time and a plethora of circumstances, which we do not yet understand, during which life might have arisen, somewhere, from the soup of elements, and led to the evolution of bacteria.

If the cradle of life is not Earth, scenarios of the spontaneous generation of life, born of carboniferous molecules and guided by RNA, then DNA, must have played out elsewhere. We should then imagine other theories and other mechanisms at the origin of life in an environment and circumstances still unknown.

With these new data, the scope of life is enlarged. Scientists, philosophers, theologians, novelists: to your pens (or processors)! The origin of life is *elsewhere.* The divine magic wand that produced life on Earth was shaped like a meteorite! Fiction—the legendary Martians—rejoins reality if we replace the little green men by tiny green bacteria.

The balance sheet of the last 50 years of research on the origin of life is simple. No empirical evidence supports the hypotheses of the spontaneous appearance of life on Earth from nothing but a molecular soup, and no significant advance in scientific knowledge leads in this direction. Even if our alchemists one day reconstruct in their laboratories part of the puzzle of how bacterial machinery arose, it will be difficult, if not impossible, to prove that that is how things actually happened on Earth. By contrast, we are witnessing the emergence of a group of new arguments and a new piece of evidence (the magnetite necklaces) in favor of an alternative hypothesis: an extraterrestrial origin. I will add that, if this hypothesis is supported by other sorts of tangible evidence, many of the articles contesting the first reports that appeared in 1996 will be thrown into question. After all, if a research team finds signs that fossil traces of bacteria are present on a meteorite and if another team proves that these traces could be of inorganic or terrestrial origin, there is no decisive reason to choose between the two hypotheses—there is doubt. The two demonstrations cancel one another. However, if other lines of research prove that the rock really does have traces of life, it is worth reexamining the previously disputed finding.

Moreover, ALH84001 has several traits that are as likely to have arisen by mineralization as by biological means. It is painful to admit that several processes that are completely independent of mineralization but that produce products resembling those that occur in living organisms could have occurred in such a small rock, even though the alternative hypothesis—a biological origin—explains all the questionable traits.

Of course, it is necessary to await other research before accepting definitively the scenario that life exists (or existed) on Mars along with all the collateral aspects of the hypothesis (that life on Earth came from elsewhere). However, at the dawn of the twenty-first century, scientists who labor to reveal the secrets of the origin of life are divided into three groups. Two small groups consist of the skeptics and the convinced, respectively. The third group is the largest: it is composed of undecided individuals who await further evidence. Therefore, whatever one thinks about them, it must be recognized that the meteorite discoveries have overthrown the age-old scientific and spiritual certainty that life arose on Earth. And it is no longer the vulgar gossip of jealous individuals or ignoramuses that must be considered in this debate. In fact, some dissidents have rushed to scorn these recent discoveries about the possible traces of life in meteorites, advancing dishonest arguments. According to them, all this publicity is just a scientific and media hype aimed at getting research funds or enhancing reputations. Whenever a scientific controversy arises, someone raises these two red flags—increased funding and reputation-building. It is easy for scientists or reporters to level this criticism, and their fame increases when they do it.

A small rock and its minuscule magnets make us dream. But if the extraterrestrial origin of life on Earth is verified, it will constitute a scientific discovery as important for humankind and for our understanding of our life on this Earth as those of Copernicus, Galileo, Lamarck, and Darwin, to which I would add those of the Nobel laureates Watson and Crick, who determined the structure of the essence of life, DNA.

Friedmann ended his tale. For a moment I thought that perhaps I had before me a Galileo or a Darwin of the twenty-first century. But he was only an ordinary man, still not well known or publicized by the media, who appreciated a French family table and ambiance. He was just one researcher among a dozen who comprise two or three leading teams that were fortunate enough to discover new facts pointing to the existence of life elsewhere. He reported to the international scientific community what he had seen and deduced. He discussed what he thought. He did not understand the hateful skepticism of colleagues who chastised him for having dared to publish conclusions that they judged to be premature. Then we changed the subject, recalling our pasts, our roots (his Hungarian and mine Dutch), his adaptation to America, mine to France, our various research subjects. But in his sensitive regard, through his fragile glasses, I recognized a certain distress, an inability to understand how he had become the object of suspicions of intellectual dishonesty. He sat there, so distinguished and calm, expressing himself in a gentle and unassuming voice, unable to bear the thought that certain scientific Yanks, to put it bluntly, had tried to bring him down by attributing to him nefarious motives that are always easy to concoct.

Friedmann again showed me the photographs from his article, wanting to reemphasize a point not sufficiently stressed in his publication. He wanted badly to convince me. He repeated his reasoning. He pleaded the correctness of his thinking. He flashed a sign of satisfaction when I asked him difficult questions that he could answer instantly. At the end of our discussion, I had to agree that his observations had convinced me and that the conclusions he drew from them had overwhelmed me. He was very likely correct.

Friedmann, a retired professor, needed calm more than controversy. Whether his discovery would ultimately be confirmed or rejected was of little importance. The scientific evidence today shows that he was right; his observations and deductions were correct and were published in a prestigious journal, they were screened by the most critical referees, and to this day they have not been contradicted.[66]

The Search for Other Evidence of Panspermia

Since Friedmann's discovery, the magnetite chains remain the best evidence for life in a Martian meteorite. Since then, many research teams have refined our knowledge of the magnetite chains, now considered a complex cellular element (organelle: magnetosome), the function of which remains to be determined. The matrix that concentrates iron to form crystals (and which surrounds each crystal)[67] and the truly remarkable magnetic fields that have been shown to exist between the crystals when they are aligned in a chain show that the arrangement of magnetites in a necklace and the regular spacing between the crystals are notable biological constants that inorganic processes could not have produced. This research reinforces Friedmann's observations and deductions: the presence of magnetite chains in ALH84001 implies the presence of bacteria on Mars 3.9 billion years ago.

As were many others, Friedmann was able to see that one encounters difficulties in being the first to describe new, upsetting evidence.

The lyrics of a song on this theme by the French singer Guy Béart illustrate this fact beautifully. It begins, "the first one to tell the truth . . . he will be executed . . . ."

This is indeed so in many circumstances for all people, including scientists.

But we are no longer in the time of Galileo or Darwin; research advances in great leaps, and all discoveries can be disseminated, uncensored, very quickly throughout the world. As the discoveries and analyses of carboniferous meteorites spurred a marked revival of interest in life on Mars, research on the topic became the most attractive objective of nations developing space exploration programs. Considerable efforts, therefore, have been and will continue to be expended in the search for traces of Martian life or bacteria. As I finish this book, the Japanese space probe Nozomi, launched toward Mars in July 1998, was to have reached the planet at the beginning of 2004 but was lost. Three other probes launched toward Mars in 2003 should increase our knowledge: Mars Express (a European probe with the robot Beagle that has remained disappointingly silent after its landing on Christmas Day 2003), and Odyssey and Global Surveyor (two American probes carrying Mars Exploration Rover robots that were well situated on Mars in January 2004). Finally, press releases from the United States announce an ambitious program: to send other probes to bring back rock samples, and finally to send a manned probe.

Even to hope to find traces of life on Mars, one must first prove that water existed there and still exists as ice. This has been demonstrated by several indirect means (e.g., traces of hydraulic erosion) and direct ones (thanks to spectral analyses and high resolution stereo photography by Mars Express from the European Space Agency and NASA's spacecraft, plus analysis of photographs by the two Mars Explorer rovers).[68] What remains is to seek traces of fossil life (on the planetary surface) or extant life (in the deep shadows of rocks where liquid water might be found).

The robots that have been examining Mars since 2004 are not equipped with apparatuses that can detect the files of magnetites that are, to this day, the only remains of bacteria able to persist in quantity at the bottom of large, ancient seas (which are not, as on Earth, roiled by tectonic movements). Rivalries between disciplines (geochemists versus microbiologists), dogmatism, and the scorn of some teams with colossal budgets with which to search for life on Mars with space probes (and who can take umbrage at the simple examination of a rock) are but temporary incidents slowing by a few years the full use of magnetic fossils to detect ancient traces of Martian life.

Research on magnetites in meteorites and on the presence of water on Mars inspire optimism that extraterrestrial bacterial forms will be found. Other recent works deal with the principal molecules of life (amino acids) forming the proteins. They were found in meteorites and in interstellar ice.[69] Research suggests clearly that the three-dimensional molecular structure of all the building blocks of life on Earth took shape in space. It has even been shown that radiation in space works disproportionately against the right-handed forms of these molecules. This is why the left-handed forms of the principal organic molecules are more common on Earth.[70]

The growing scope of methods used by astrobiologists and the spectacular growth of scientific communication and competition on this subject are strong signs of progress in considering the possibility of an extraterrestrial origin of life. I am convinced that definitive signs will be revealed by the various current research approaches. It is still necessary, however, to convince the public of the validity of a hypothesis that seemed heretic in the twentieth century. Life came from *elsewhere*. It will take time before this idea finds its way into textbooks without reservation.

I am ending this chapter on the birth of an intellectual revolution. Recent scientific advances buttress the hypothesis of an extraterrestrial genesis of life. With the observation of magnetite chains in ALH84001, we are forced to open our eyes and realize the enormity of what this modest discovery suggests. *The seeds of life could have arrived on Earth thanks to introduced species that became naturalized and proliferated*. This hypothesis is presently better supported than the alternative, which until recently seemed obvious—that of a terrestrial origin from inorganic matter, *ex nihilo*.

We can only wait for other lines of evidence. If they support Friedmann and other pioneers of astrobiology, it will recall the slow emergence of Copernicus's hypothesis (the Earth revolves around the sun) and Lamarck's (humans derive from apes). It will then be exciting to follow the stages of common acceptance of the extraterrestrial hypothesis of the arrival of life on Earth. By contrast, if new evidence disproves the existence of traces of life on Mars, this entire story will be just an exciting episode in research on the roots of life. We can bet it will not be the last.

Whatever happens, the first traces of life found on Earth, which existed more than 3.5 billion years ago, are definitely bacteria and were very diverse even then. We will see in the next chapter that they reigned alone on Earth for almost a billion years and that they still dominate life on Earth today.

CHAPTER 3

Papa, What's A Bacterium?

On the First Life on Earth

One day at Bible school my daughter Coralie, who was then 12 years old, heard the minister mention Genesis, the first pages of the Old Testament. This remarkable story includes all the elements of the puzzle of life that God established. The followers of many religions (including Christianity, Judaism, and Islam) believe that these writers of the account of the origin of life, who recorded the stories at a time when knowledge was rudimentary, had surely received divine inspiration. While supporting this mystical description of the origin of life, I sketched out the scientific hypotheses about the appearance of life on Earth—I told her about bacteria.

"However life on Earth originated, the first organisms known to have colonized our planet were bacteria."

Coralie is mischievous. She tried a diversion to avoid my long, boring tirades: "Papa, what's a bacterium?"

"Uh-oh." A simple enough question, but the answer is not so easy. I threw the question back at her: "What do you think it is?"

"Hmmm." And after a little hesitation: "A tiny thing, a germ. Germs are bad things that make us sick. We shouldn't have germs on our bodies; that's why I wash my hands before eating."

"Not bad!" I understood at that point how abstract the question was. And as she had succeeded in distracting me, I left her comfortable with the image of Genesis that the minister must have presented so well.

"I'll explain it some other time."

Subsequently, I conducted a little poll among the friends of my adolescent children. To the question, "What's a bacterium?," the responses varied from "I could give a damn!" to descriptions of present-day illnesses caused by bacteria.

I understood that, for many, the world of bacteria was an unreal world. I deduced that for the majority of non-biologists, the term at best suggests invisible, unhealthy germs. However, ever since Antoni van Leeuwenhoek observed them in the seventeenth century and the French biologist Louis Pasteur (1822–1895) recognized them as germs that did not arise spontaneously from nothing, a wealth of research has elucidated the nature of bacteria.

How to interest people in this subject? One summer day my family went swimming at a public beach. The idea struck me to show Coralie the notice about the health inspection of the water that was posted at the entrance. She was already in the habit of consulting the announcement board to learn the water temperature and whether jellyfish had been spotted. She was afraid of jellyfish; the presence of just one of these gelatinous, stinging monsters sufficed to dissuade her from swimming. The health inspection notice, which had been posted three days earlier, noted "high-quality water," and there was an index, standardized throughout Europe, which signified the same thing. I showed her this mysterious evaluation. How could one assess the "quality" of swimming water?

I explained to her that this assessment referred to pollution primarily caused by sewer water that, more or less purified, is poured into the sea, along with its load of bacteria that could infect the skin, ears, or eyes. To assess this pollution, one simply counted the number of several species of bacteria chosen to represent all those present in waste water that are dangerous to our health. The number of bacteria per liter of water is related to the risk that we will get sick.

"Technicians came here, sampled the seawater, and counted the bacteria: *Escherichia coli* and other coliforms, *Staphylococcus,* and fecal *Streptococcus*. Three days ago, there were few of them in this water—not enough to be dangerous. That's reassuring. It reassures swimmers—like you, when you learn there are no jellyfish."

I waited for her reaction.

"But how can they see them and count them, and why didn't they come this morning to sample the water?"

I then described a period I had spent in a public health laboratory when I was a student. For three months, I had sampled and analyzed thousands of liters of seawater.

I began by explaining to her that one does not see bacteria with the naked eye, because they are too small. Most measure about a thousandth of a millimeter: imagine a millimeter of your ruler, cut into a thousand equal parts. A thousand bacteria lined up in a row measure a millimeter.

"With an optical microscope, the kind you use in school in biology courses, at the highest magnification, you see only little moving dots. You can't even distinguish them or count them accurately. However, you can see them really well with an electron microscope of the sort used in universities, but before you get to that point, there is a laborious preparation and you still can't count them all."

"Then how did you do it?"

"It's very simple. I took a tenth of a liter of seawater randomly in front of the beach just beneath the surface. In the laboratory, I filtered the seawater by running the water with the invisible bacteria through a circle of filter paper with the diameter of a teacup. The water passes through the filter, but the bacteria remain on the paper—the bad ones that we are looking for, and the others.

"You should know that, on average, there are more than a million 'natural' bacteria, not caused by pollution, in a single drop of water!

"Then I placed the filter paper in a container that has a sort of jelly that the species of bacteria I was trying to count really like. After I covered it, I put the container in a warm place that had the temperature most favored by the kind of bacteria I was looking for. You have to do all this very carefully. The filter paper, the jelly, and the container all have to be disinfected and the manipulations must be done under a sterile hood to avoid contamination by bacteria from other sources. Finally, you wait one or two days, depending on the species. During this time, the invisible bacteria on the paper multiply. They do it so well that from every individual bacterium you get a mass, a colony that you can see with the naked eye. Only the colonies of bacteria harmful to our health have grown, because they were the ones we selected for by using certain culture conditions. Then, all you have to do is to count the colonies."

"How do you do that? You're going too fast."

"It's very simple. Each colony arose from a single bacterium that was in the water. So by counting colonies, you get the number of bacteria that had been in the quantity of water that you had sampled and filtered two days earlier. The results are then interpreted and confirmed by two or three officials before being posted. Altogether, this usually takes three days. That explains the delay in getting the information to the public."

"OK, I understand how you count them, but it takes a long time and . . . well, maybe today the water is really bad and nobody knows it."

"Exactly! But we don't know how to do it any other way. But calm down—when the water from some beach is often bad, they prohibit swimming and try to improve water purification in the area."

As Coralie finally understood, bacteria—microscopic, some harmful, and some harmless—swarm in seawater. But they are not found only there; bacteria reign over our planet. Since the day Coralie learned this lesson, I have often cited this personal incident to show that one can see bacteria by simple techniques, making them more concrete, more "present."

On the beach, we were sitting side by side, watching the sea and the bathers. We warmed up a bit before going into the water.

I launched into a description of the invisible world of bacteria. I no longer remember if she listened to me, and I do not remember too well exactly what I said.

I must have begun at the beginning.

Bacteria Adapted to Many Lifestyles

Between the first minuscule bacteria that left a trace on Earth and the present day, more than 3.5 billion years have passed.[1] For nearly one-third of the time life has existed on our planet, bacteria reigned alone. There were only microbes of different shapes and functions. How many species? We don't know, we will never know.

Right at the beginning, there was no organic matter from which the bacteria best known today could have nourished themselves—that is, all those that sustain themselves as fungi do by dissolving matter constructed by other organisms.

Therefore, the first organisms that were able to grow on Earth in abundance were bacteria able to live in an environment of hot liquids coming from the heart of our planet or other extreme habitats. In these environments, these bacteria assimilated inorganic matter by chemical reactions in liquids and gases without oxygen. I will give them a simple name that well suits their lifestyle based on chemical reactions: *chemobacteria*.[2]

We still find such species today. They are abundant in astonishing conditions. Some are found in very hot water (above 80°C).[3] Among these, some can survive at 113°C, as long as they are under sufficient pressure to keep the water from boiling.[4] We find these bacteria around hot-water geysers and in the depths of ocean faults, where plumes of scalding, sulfurous mud are exhaled by the Earth.

Other bacteria thrive in water at the freezing point and some grow under ice in the frozen Antarctic deserts where the mean temperature is about −68°C.[5] Some bacteria can survive being marinated in solutions with acidity comparable to that of sulfuric acid,[6] while still others proliferate in hypersaline water (brine), where they live on the verge of crystallization.[7] Finally, other types assimilate carbon dioxide, using hydrogen as an energy source; they excrete methane, which we use for cooking and heating.[8] All these bacteria have reputations as organisms of extreme environments;[9] they would be far in the lead in many chapters of the *Guinness World Records*.

These last two decades, discoveries of chemobacteria buried deep under our feet have multiplied. They have been found up to 10 kilometers underground in the heart of hot, humid rocks. As energy, they use hydrogen derived from a chemical reaction that modifies rocks in contact with hot water under pressure.[10] These bacteria—invisible, living in rarely visited habitats of perpetual darkness—are so numerous that they form an ensemble called SLiME.[11] This bit of British humor is an acronym for Subsurface Lithoautotrophic Microbial Ecosystem (a simple translation: ecosystem composed of microbes that eat subsurface rocks).

The American geophysicist Thomas Gold proposed a measure of the volume of the habitat for this microscopic life existing below the soil. For the depth of the habitat, he used 5 kilometers (although bacteria can live up to 10 km below the surface). The surface area was that of the entire Earth, including the entire substratum under the seas and oceans (which cover more than two-thirds of our planet). He multiplied the thickness by the surface area to obtain the volume of favorable habitat for the bacteria of SLiME. Gold estimated that the space occupied by pores and microcavities that could contain the bacteria was 3% of these pores. Finally, he estimated that bacteria actually occupy 1% of this volume. The result was surprising: the volume of chemobacteria would form a layer 1.5 meters thick covering all dry land on our planet.[12] A viscous layer of life greater than the volume of the entire fauna (humans included) and flora (including tree trunks) mixed into a mush and spread over the same emergent land!

At a stroke, by a simple calculation, life assumes another aspect. It is no longer a dense tropical forest that best represents the voluminous exuberance of life on Earth, but a blanket of invisible bacteria that live under our feet, eating rocks.

The more we learn of these chemobacteria, the more we think about the hypothesis of extraterrestrial life. These organisms, growing in the least propitious conditions, have been able, and would still be able, to form hidden populations at the heart of certain planets. To live, they need only water and rocks. In our solar system, Earth provides these conditions, but it is likely that Mars, Venus, a satellite of Saturn (Enceladus), and three satellites of Jupiter (Europa, Callisto, and Ganymede) have the necessary environment, where water is likely to be frozen, evaporated, or infiltrated into the ground. It was with the primary objective of collecting data on the presence of water and life—and the hope of finding traces of living or dead chemobacteria—that the United States, Europe, and Japan launched probes towards Mars in 2003.

Very early in the history of life on Earth, between 3.5 and 2.5 billion years ago,[13] other families of bacteria, which had very different lifestyles, cohabited in water with these rock-eaters. Their main trait was the possession of pigments that served to capture and transform light energy. The most extraordinary pigment, in terms of its complexity and function, is chlorophyll. It was "invented" by these bacteria. It is exactly the same chemical that one finds today in lettuce and other green plants. This molecule transforms solar energy in order to assimilate inorganic matter (the light energy arrives in the form of particles called photons—hence the term "photosynthesis" to designate the process of synthesis of organic matter thanks to light). This reaction produces oxygen, which dissolves in the water. These plant-like bacteria[14] then began to oxygenate the water they lived in. In that era, oxygen would have been considered a waste product of the functioning of plant-like bacteria with chlorophyll. This is how the first global change occurred. Over several hundred million years, imperceptibly at first, then ever more quickly, the liquid environment of life (the hydrosphere) changed. The upper part of seas and oceans, where the plant-like bacteria lived, became rich in oxygen, a pollutant for the other bacteria. As millennia passed, oxygen became highly corrosive in the upper meters of water, killing sensitive bacteria for whom this gas was deleterious.

In this way, what we dread these days—a significant global change—has already happened. A dominant group of organisms (the plant-like bacteria) poisoned everything with its waste. The difference between what happened then and the current situation, where man is changing the atmosphere by generating waste gases that produce a greenhouse effect, is that the prehistoric global change was very slow. In effect, from the known dawn of life on Earth, where plant-like bacteria were already present, they took hundreds of millions of years to poison other bacteria with their gaseous waste. By contrast, current modification of atmospheric gases is racing, in terms of the geological time scale; for the last two centuries, significant changes can be seen on a scale of decades in the increasing concentration of greenhouse gases.

Certain bacteria became increasingly adept at tolerating oxygen and then using it while remaining in the upper waters of the oceans. Others could not handle the change and remained buried in the mud, sheltered from the oxidizing gas. The groups that opted for these two strategies are still present today—descendants of bacteria that either escaped or adapted to the first global change.

The flora and fauna that we know today cannot live without oxygen and could never have evolved without the labors of these plant-like bacteria, which still exist.

A third type of bacterium, some adapted to the water oxygenated by those already discussed, and others, not, subsequently appeared on Earth. These were specialists in digesting organic matter that the entire community of bacteria began to accumulate en masse. They found energy and nourishment assimilating the wastes and cadavers of other bacteria (even in the form of molecules dissolved in the water). We call them "fungal bacteria" because, like fungi, they absorb organic matter by simple contact.[15] Much later in the history of the development of life on Earth, beginning 570 million years ago, when many plants and animals visible to the eye had begun to proliferate in the oceans and then on land, these bacteria evolved to live with these new organisms, for better or for worse. Some continued to break down inert organic matter (excrement, animal cadavers, dead algae, leaves, and wood). Others became parasites, living to the detriment of living animals and plants. Many of these are very well known; they cause infectious diseases (anthrax, leprosy, tuberculosis, cholera, plague, diphtheria, smallpox, tetanus, syphilis, Legionnaires' disease, gastroenteritis, septicemia, abscesses, some forms of pneumonia, sore throats, meningitis, and many others). There are also parasites of plants or of our cells that resemble minuscule fungal filaments.[16] Finally, other fungal bacteria are "friends"—including those that allow the transformation of milk into cheese or yogurt and those that swarm in our guts and help us to digest (known by the reassuring term "intestinal flora"), or those that are associated with plants and help them fix nitrogen from the air, which they transform into ammonia (clover is such a plant).

These bacteria have highly varied shapes: they can resemble spheres or spheroids, rods or cylinders, and may be equipped with a whip-like appendage. In some species, the rods are aligned end-to-end. Other bacteria live in dense colonies, forming a jelly-like, undifferentiated mass. Still others are arranged side by side to form true filaments, branched or not, producing colonies visible to the naked eye (fig. 5). Their colors (discernible in species producing large filaments or when the microscopic cells form colonies of hundreds of millions of individuals) can be black, white, gold, blue, green, or red. Cell size varies from 0.0002 to 0.66 millimeters wide.

Today, the bacteria that have been identified and baptized with the name of a genus or species number fewer than 3,000.[17] This small number is explained by the very formal rules for identification and naming. In fact, a species of bacterium cannot be named until it has been cultured by several laboratories. The stock, isolated and named, is therefore available to the scientific community for evaluation, study, and experimentation. This is the only way to allow for comparison of a new type of bacterium with those named and available in cultures. However, the great majority of bacteria cannot be cultured and so remain nameless. Several methods of genetic analysis of the bacterial community living in diverse habitats (soil, humus, marine mud, living organisms) lead to the conclusion that an immense number of diverse bacterial species remain to be discovered and identified.[18] Millions of species or stocks of different bacteria must exist today on our planet.[19] Each constitutes an original chemical and biological system constructed in the heart of a tiny cell, piloted by specific genetic codes. To attempt to distinguish them, even in imperfect fashion, researchers have begun to inventory them even when they cannot culture them. They give them code numbers corresponding to the DNA sequence of one or several of their genes. These sequence descriptions are deposited in international banks of genetic data accessible to anyone.

Bacteria Visible to the Naked Eye

Microscopic and mostly invisible, these numerous and diverse microbes escape our attention. An abstract world, even one so important for life, does not arouse much attention. Coralie had tuned me out; she no longer listened to my explanations. She stared at the sea, the sailboats in the distance, the movements and forms of the swimmers.

I then concocted a little diversion. Taking her hand, I proposed going to collect bacteria.

"You're going to see and touch them! Come on over here to these rocks."

Incredulous but curious, she followed me. On the rocks arranged to form a protective jetty for the beach, we could easily discern, where waves

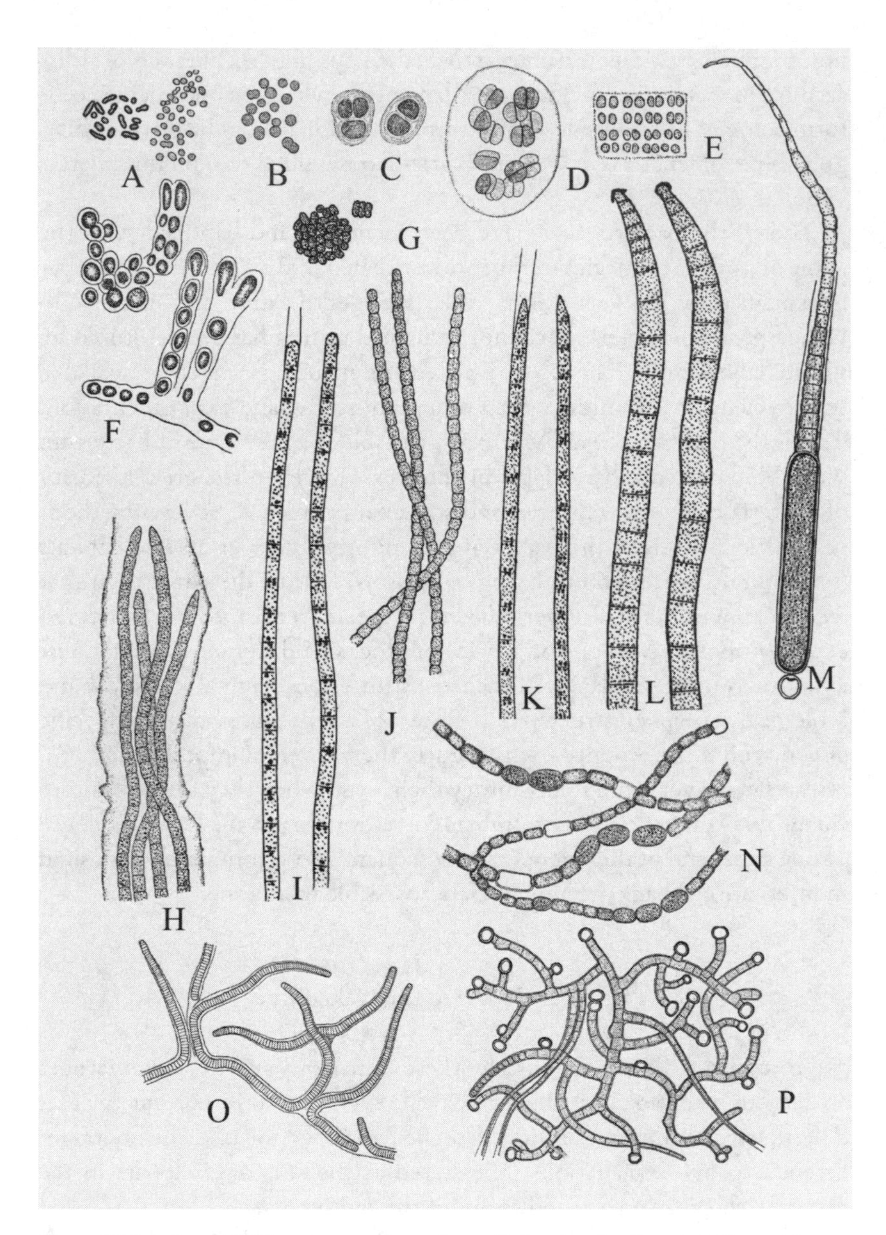

Figure 5. Drawings of different types of cyanobacteria showing the diversity of bacterial forms: unicellular (*A*, *B*), colonial (*C–G*), multicellular (*H–P*), and cellular differentiation (*L*, *M*, *N*, and *P*). Note the different cells: akinetes and heterocysts. (Drawings by Pierre Fremy.)

were washing up, a colored fringe of algae of diverse shapes, especially brown and green algae. With a nail, I lifted to her eye level small hemispherical, blue-green piles.[20]

"They look like little half-peas!" she exclaimed.

"They're bacteria, colonies of filamentous bacteria that contain chlorophyll. They function like plants: they're plant-like bacteria."

Later, I found wine-colored filaments.[21] These hairy algae are also plant-like bacteria, with more red pigments than the other species. Coralie looked with astonishment at these concrete, harmless bacteria that rested in the palm of her hand.

These are representatives of one group of plant-like bacteria. For many years they were considered to be types of algae—plants with a structure that is often rudimentary but in which the cells are much more elaborated and differentiated than those of bacteria (in particular, algae have a nucleus containing the chromosomes).

The genes of these plant-like bacteria nevertheless contain all the information necessary to organize the construction of filaments comprising several cells lined up side by side, sometimes cells of different types; these are cases of cellular differentiation (fig. 5).

"If we were in Holland, at the edge of lagoons of the North Sea, I would have been able to show you plant-like bacteria that creep. They live on the mud and are made of filaments twisted like a corkscrew.[22] At high tide, they are submerged and they cover the mud, which has a greenish tint. At low tide, they are out of the water and are exposed to the vagaries of the weather. To protect themselves, by contraction, they 'creep' under the mud and disappear temporarily. The mud then appears to be beige or gray. You see, bacteria can be everywhere, in forms and places you would not have expected."

She rushed to show the colored, slightly viscous bacteria to her big sister Marjorie, who was more disgusted than interested.

"Papa already showed it to me. Get rid of those slimy things and let's go swim!" Marjorie took Coralie to shield her from the tirades of an old, daydreaming professor. I smiled because I knew that, from all my explanations, they retained something.

The Slow Colonization of the Earth by Bacteria, and Bacterial Evolution

Facing the sea, I took up the thread of my thoughts again and imagined the scenario of the colonization of our planet by bacteria. The first bacteria that appeared on Earth must have arisen in a particularly favorable spot. If they came from elsewhere, on a meteorite, they had to have landed in just one place. To conquer all the areas that were favorable at that time from the point of landing must have taken a long time. And why not two meteorites, or several, all carrying life and landing in various places?

Whichever scenario is correct, the first bacteria on Earth multiplied, then dispersed; by 3.5 billion years ago, they had reached many regions of the planet, as shown by their fossilized vestiges, which are well scattered.[23]

In the water, these minuscule cells can float and be carried by currents. They can also insinuate themselves in the tiniest interstices of the humid, hot terrestrial crust. In the air, the wind can carry them great distances. Much later, the animals they have parasitized or with whom they are associated would also turn out to be a good means of transport. Bacteria have therefore been able to colonize volumes and spaces much greater than those where animals and plants are currently found.

It was by contact with many environments and circumstances of life encountered during these voyages that bacteria diversified.

It is a fact: the extraordinary diversity of bacteria is tied to their minuscule size, which favors their dissemination. This mobility allowed the isolation of particular stocks, after which natural selection of the most resistant and most prolific types occurred in various environments. These adaptations to extremely varied conditions, acquired through evolution, have been induced by a remarkable potential for modification of the cell. They are still changing today. The principles of evolution apply to bacteria as to all living organisms.

Evolution is a dynamic process. Time is needed for a modification to become established and to perpetuate itself. These organisms have persisted for millions of millennia, evolving internally while not changing their size.

To evolve, it is necessary to reproduce. Most often, bacteria reproduce very simply: they cut themselves in two. Genes and other matter are duplicated and equitably shared. In this case, there is no sexual reproduction. The duplication of cells is not followed by a fusion of a male bacterium with a female bacterium. There is no interbreeding, no hybridization; this is a disadvantage for their evolution. In fact, the handicap of their very simple reproductive mode is compensated for by the speed of their doubling. In one hour, several divisions give several generations. During each division, minuscule errors can appear. These copying errors are mutations at the genetic level. Therefore, the larger the number of divisions, the greater the number of errors. And the stress of changing environments can favor errors in transcription of genes during their duplication. Knowing that certain bacteria divide in less than half an hour and that this phenomenon was initiated on earth more than 3.5 billion years ago, we see that there have been a gigantic number of modifications. So this system of simple duplication can produce a huge number of potentially innovative genetic changes. This system, favorable to adaptation to new environments, based on copying errors, is complemented by a gene exchange. Bacteria can in fact transfer genes to one another without true sexual reproduction, most often with a virus as intermediary.[24]

All these transfers, exchanges, or modifications of genes lead to the expression of small changes of form or behavior. These changes then confront the environment in which the bacteria live. Only the changes that are viable—advantageous or at least neutral—are perpetuated.

A classic example illustrates the great potential for transformation of bacteria. You have a sore throat caused by bacteria, so the physician prescribes medication (an antibiotic) to kill them. You feel better after two days and stop taking the medication, contrary to the directions. This is a mistake, because, among the billions of bacteria that have proliferated in your throat, some are more resistant because of infinitesimal differences in their genetic code. No longer having to face the antibiotic, the surviving bacteria, different from those that have perished, are going to proliferate, and the antibiotic will be less effective against them. This is how excessive and repetitive use of antibiotics leads to the emergence of new pathogenic strains of bacteria against which existing antibiotics are useless, so new medications must be invented. These, in turn, are likewise often used inappropriately and lead to the appearance of still other resistant strains.

This emergence of resistant strains is so quick that scientists have invoked the idea of inheritance of acquired characteristics in some cases. This suggestion arises from the hypothesis that, in the course of the short existence of a bacterium, it has somehow "learned" to defend itself. This "knowledge," in turn, somehow modified its genetic code. Such bacteria thus transmit to their descendants an acquired advantage: this is the mechanism of evolution proposed by Lamarck.[25]

Whatever the truth in the above cases, the fact remains that the immense diversity of bacterial species and of their modes of life results from the combination of two characteristics: their small size (favoring their dissemination to very varied environments) and their rapid division (engendering modifications of their genes). Bacteria are therefore opportunistic voyagers, possessing a mechanism of automatic genetic modification that allows them to change, to innovate endlessly, and therefore to adapt quickly to new conditions.

The astonishing diversity of bacterial modes of life leads to questions: Has a chemobacterium been able to evolve into a plant-like bacterium or a fungal bacterium? Is simple evolution at the heart of bacteria? In which direction? Or was the Earth inoculated with several different bacterial strains?

Nowadays, several hundred researchers devote their lives to generating hypotheses and solving these mysteries. Because representatives of these types of bacteria still survive, their genomes can be studied. We thereby find features common to these three groups (a similarity that was already evident, because all bacteria have the same internal components). Scientists have also studied their degree of relationship by examining the greatest genetic divergences or convergences. These studies have led to a surprising observation: the majority of chemobacteria are very different from all other bacteria. This group of chemobacteria, often living in the most extreme environments, has been named *archaebacteria;* the others, all the others, are the *eubacteria.*[26] Therefore, the ensemble of bacteria can be divided into two distinct trunks of a genealogical tree (specialists in classifying organisms use the term *kingdom*) representing two types of biological factories constructed on the same basic plan but with different motors and materials.[27] But the two kingdoms, or evolutionary trunks, are so close that small sequences of genetic information are identical. These may be the heritage of exchanges, of transfers, or of inoculations of genes by viruses. Geneticists think that the two kingdoms of life arose from the same base: archaebacteria and eubacteria must have had common ancestors. But geneticists' abilities have limits, and so the debate remains unresolved as to when, under what circumstances, and where the common basis of these two distinct lineages was formed.

Among the bacteria that currently exist on Earth, if certain ones have changed little or not at all in 3.5 billion years (such as those that inhabit stable environments like humid, warm, rocky crevices), others have evolved endlessly. This is especially true of those that have evolved to live at the expense of the millions of macroscopic species, multicellular organisms that comprise the flora and fauna that appeared relatively late on our planet (about 570 million years ago). An example is the majority of pathogenic bacteria that we fear; they are specific to our species and cannot live except to the detriment of humans. These species thus arose with or after the appearance of the predecessors of humans, so they have existed for at most several million years. Before them there were other bacteria, as before us there were other primates.

Knowledge of bacteria leads us to understand that the evolution of life is not fated always to produce complex organisms.[28] Evolution need not systematically lead to multicellular, large, intelligent species. Among bacteria, more than 3.5 billion years of evolution, often very rapid, has led them to remain minuscule, though very diversified in the way they function, and spread all over the Earth. Adaptation to such diverse habitats has been internal, achieved by modification of their cellular biochemical factory in the interior of their isolated, microscopic cells.

Plants and animals are the result of other evolutionary pathways, other geneses. They arose very late in the history of life and hatched amid the enormous kingdom of bacteria.

Bacteria: Champions among Life Forms (All Categories)

In considering what I have just written, I must conclude that the connotation commonly given to the word *evolved* arises from the fact that man is taken as a standard. In effect, it seems to us that we are the most highly evolved living organisms. This is why the most widely known conception of evolution rests on a scale of values with us as the reference. Even so, man belies this connotation.

To rank degrees of evolution we refer to the size of organisms (but if this criterion were used alone, dinosaurs, whales, and elephants would be more evolved than we are). We also consider the structure of the organism, assuming that ours is the most complex, the most highly evolved (although that of other mammals is similar). We refer above all to our intelligence compared with that of other forms of life. In this sense, we concede that certain familiar mammals (including chimpanzees, dogs, horses, and porpoises), certain birds (like parakeets), and even certain invertebrates (such as octopi) have behaviors that we consider intelligent. We feel they have attained a certain degree of evolution, though it is very inferior to our own.

In fact, if we consider other criteria in determining evolutionary rank, we obtain a different order. If, for example, we include the time that has passed since the first appearance of a lineage of organisms and the number of generations and changes that have led members of that lineage to be adapted today to different environments, bacteria undoubtedly constitute the most highly evolved lineage of organisms on Earth.

The word, *evolved,* used to illustrate our apparent superiority over other life forms, along with our self-consciousness, also corresponds to a deep-seated ideological and philosophical image of domination. Man as dominant is a generally accepted idea, the product of the Judeo-Christian and Muslim religions that depict us as above all other forms of life. We believe that we have achieved the objective God assigned to us in Genesis: we dominate the fauna and flora. Our social organization, our numbers, and the power of our technology also confirm our impression of our dominion over Earth. However, bacteria are doubtless the truly dominant organisms on our planet:

- by the area and volume they occupy
- by their global mass
- by their number[29]
- by the population of each species[30]

Better constructed to survive all the cataclysms, able to live in slow motion while waiting for better conditions (like those bacteria resuscitated after resting 250 million years in a drop of water trapped in a salt crystal[31]), and highly varied in their modes of nutrition (their nutritive resources are infinite), bacteria have all the assets needed to persist on Earth indefinitely.

If they knew of their lowly status here, they would have no existential fear. Their populations are in equilibrium with the environments they have inhabited for hundreds or thousands of millions of years, and their descendent generations will never confront limitations imposed by a self-destructive development. Thanks to their extreme diversity and to their rapid evolution, there will always be bacteria who will find a way to live on Earth for several billions of years into the future, until the extinction or explosion of our sun.

The dominance of bacteria should also throw into question our feeling of invincibility. If a deluge of nuclear bombs irradiates the entire Earth, if a giant meteorite strikes our planet, or if some stellar phenomenon changes even slightly the position of the Earth in its solar orbit, our civilization, social structure, intelligence, self-consciousness, and knowledge will be of no avail. We will disappear forever. By contrast, the bacteria will resist; their survivors will be able to continue to multiply, to diversify, to spread, and—why not?—to evolve other floras and faunas.

And let us not forget that our lives also depend on bacteria that live in our bodies. By the same token, our lives are also at the mercy of a new pathogenic bacterium, more opportunistic than the others, that will find in man a paradise to proliferate in and to annihilate.

Finally, the abstract kingdom of the invisible world of bacteria is a magnificent subject for meditation on the absence of a predetermined direction of evolution toward complexity or progress (to remain minuscule is the preponderant path of evolution, because small size is favorable to growth in all terrestrial habitats), and on our humble and uncertain status on Earth, where we are threatened by bacteria, which are assured ultimately to survive our disappearance.

We therefore should not seek to distinguish superior or inferior species in the biosphere. There is not progress in some and regression in others. There are not developed species and underdeveloped species. Each species has a distinct evolutionary history and a place among the other species in a habitat where it can thrive.

I came to my senses on the beach again, regarding the blue of the sea and rolling a pebble from one hand to the other. I thought of certain Greek philosophers who had just deduced that their mythic gods did not exist and that man was indeed alone on Earth, confronting his destiny. If they were beside me today, their thoughts would go towards that invisible world of lives that silently dominate our planet. Man is still alone facing his destiny, but he should understand today that he does not constitute the dominant life on this planet. He is only a fragile element of life, and his fate is at the mercy of a minor astronomical event or of a parasitic bacterium more pathogenic than the rest. His destiny depends on other lives that surround him and, recently, on his impact on them.

I remember coming to my senses and also being conscious of my physical surroundings. A beautiful woman with bare breasts arose and went into the water. Coralie and Marjorie came up onto the beach smiling and splashing drops of water. I remember saying to them:

"I hope you weren't afraid of bacteria."

Then: "Did you see any jellyfish?"

They had not seen any. I responded with a big smile: "Good! Then I'm going swimming."

I had just enough time to don my mask and snorkel, to jump into the water, and to swim several meters.

After having evoked the impressive evolution and power of the bacterial world that surrounds us and that is most profoundly associated with us, I must address another fundamental step in the development of life on Earth. This is the astonishing transition between bacteria and plants or animals. That is another genesis.

CHAPTER 4

The Vampire Slug of the Killer Alga

The Genesis of Unicellular Animals and Plants: The Evolutionary Revolution

The auditorium is full of young students taking freshman biology, proud to be at a university and with dreams of becoming high-level professionals. The chatting of 150 young men and women creates background noise. Below them, before a pair of immense blackboards, I feign consulting my notes while waiting for latecomers to sit down.

When the students calm down and are clearly paying attention, I ask in a sonorous voice:

"Define a plant as distinct from an animal."

I wait silently, looking at several random students, catching their eyes. This immediately intimidates them. It is not normal to begin a course by a question. Many students already take me for a character. Sensing a trap, no one answers. I make fun of the utter silence. Appearing severe and annoyed, I exclaim:

"What? You're in freshman biology and you're unable to distinguish a plant from an animal?"

The least intimidated ventures:

"But look, sir, we all know plants are green, have chlorophyll, and live by photosynthesizing, and animals don't!"

"You're correct. But what about fungi, do you classify them as animals?"[1]

Laughter in the auditorium.

"I want a better definition."

Again a long silence. Taking a chance, several students raise their hands. I call on a girl who seems determined and confident in her response.

"Plants are stationary, and animals aren't!"

"Not bad."

Several mocking whistles are audible. The student blushes. Some students pick up their pens. I don't give them time to write.

"If I am following you correctly, Miss, you would class stationary sponges and corals as plants. And thousands of species of microscopic, single-cell algae that swim thanks to their flagella, are they animals?"

In the auditorium, the students no longer laugh; the silence is deafening. They begin to question what they really know. Having been burned, they no longer want to speak. Then I give a clue.

"What would you say about this definition: plants have cells with a rigid cell wall composed of cellulose, and animals do not?"

Murmurs, nods of acquiescence; the students take their pens and write it down. I let them do it; then I intervene.

"Unfortunately, that's not good enough. There are too many exceptions. Many fungi don't have cellulose walls; they are replaced by chitin, the substance that constitutes the carapace of crustaceans and insects. And some animals construct cellulose coats."[2]

The tension is palpable. Some erase the bad definition, others cross it out. It's time to cut short the suspense and tell them the right answer.

"In fact, it is impossible to tell an animal from a plant with a simple definition. You have to refer to their origins and know that all groups of species that have cells with nuclei (all the animals and plants) are in fact composed of several distinct lineages that must be defined independently.

"It's still true that these lineages can be crudely classed among plants or animals if we consider one or a few simple traits, with some exceptions that I've just mentioned. Flora and fauna, plant and animal, indefinable terms, will always remain elementary, simple distinctions for the lay public. So we'll continue to use these terms, knowing their distinctive features and their history."

The Principle of Symbiosis: Intimate Association of Two Organisms

Then I began my lecture on the origins of the "fauna" and "flora." I need to explain how plants and animals arose from bacteria, very simple microorganisms. The mystery of their genesis is now largely solved. It is a magnificent natural history. To explain better the novel mechanism I will have to refer to, I will begin with a digression. I will recount a personal experience.

Facing the inexorable invasion of the tropical alga, *Caulerpa taxifolia,* introduced to the Mediterranean, I had a lot of trouble getting myself heard. Overshadowed by a heated controversy on the source of the algal introduction, my warnings were not heeded and no strategy to control the invasion was elaborated until it was too late to stop it.[3]

When physical and chemical means can no longer be used to reduce or control the invasion of a species, one must search in its native region for a natural enemy, like a predator, that focuses its attack on the invader. This is the fundamental condition of success: there should be no possibility of introducing an enemy that could instead attack other species, native ones.

This "biological control" is often used in agriculture (for example, lady beetles are used against aphids). Biological control has succeeded in choking off several invasions that had transformed whole landscapes, like that of the South American prickly pear cactus. Though it had invaded more than 24 million hectares in Australia by 1926, it was reduced to small, isolated populations by a small moth whose caterpillar eats only this cactus.[4] But biological control is not without risks: the same moth, introduced to the Lesser Antilles to destroy invasive cacti, reached Florida, from which it has advanced as far as South Carolina and Alabama, devouring native cacti.[5] Now biologists fear it will reach the southwestern United States and Mexico, regions where cacti are symbols of the land as well as valuable crops.

For the invading alga, the only animals known to eat it are tropical sea slugs (plate 5). It was necessary to prove that their introduction to the Mediterranean would not be dangerous and that the sea slugs could not change their diet and start eating other algae.[6]

This is why we studied these sea slugs.

Initially, it was very easy to observe that the selected sea slug—a species of *Elysia*—enveloped parts of the alga with its mouth in order to implant a sharp rasp, shaped like a row of teeth. Through the opening it has just made, the sea slug sucks the body fluid, or "sap," of *Caulerpa taxifolia* (its cytoplasm), thanks to its very muscular pharynx. The high viscosity of the fluid and the reaction of the alga (it plugs the wound with a mass of coagulated matter) force the sea slug to repeat the piercing operation frequently. To this end, the sea slug slides over several millimeters of algal surface to forage anew.

In several hours, a frond of *Caulerpa taxifolia* is thus punctured several dozen times and its cytoplasm sucked out. The frond, evacuated and spotted with whitish haloes, becomes flaccid and dies without resprouting.

This sea slug, a vampire of the killer alga, cannot eat other algae. Either the algal wall is too thick or the diameter of the veins of the algal fronds is such that the fronds do not fit into its mouth. But above all, the anatomy of the other algae is not at all compatible with the sea slug's system of ingestion by aspiration. In fact, all marine flowering plants and more than 95 percent of the algae found near shore are multicellular. Their architecture is based on the same principle as ours: a genetically programmed stacking of tiny cells of various forms. The cytoplasm of these microscopic cells cannot be sucked by the sea slugs. By contrast, in species of the order Caulerpales (including species of *Caulerpa*), individual plants are constituted of a single giant cell. This cell is basically a tube in the shape of the alga. Inside the tube, all the liquid (cytoplasm) circulates freely from one end to the other. This is the reason these sea slugs do not eat *Caulerpa* the way a vegetarian would. They do not act at all as we do when we eat lettuce. They do not bite the alga, but rather, they pierce it and suck the cytoplasm the way a mosquito "bites" us and sucks our blood. This is a very advanced adaptation that must have arisen hundreds of thousands of years ago, if not millions. To evolve different eating styles takes a lot of time.

For this reason alone, we had good reason to reassure people who feared these sea slugs would eat other algae. But this remarkable specialization to exploit one type of food is not the only unusual feature of these sea slugs.

In their quest for *Caulerpa,* sea slugs have to cross hostile terrain where dozens of predators can devour them. In theory, fish as well as crustaceans and other invertebrates could swallow them. When the sea slugs move to hunt for *Caulerpa,* they are at great risk. Traveling long distances, they would seem to be easy prey. In fact, they are never eaten because, by sucking the cytoplasm of *Caulerpa,* these sea slugs swallow toxins produced by these algae. Not only do they resist the effects of the toxins, they sequester them. This is how the vampires of *Caulerpa* become poisonous for marine carnivores. The animals know this and leave them alone. We have performed aquarium experiments with several Mediterranean fish species, starved for several days. Their initial reflex is to jump at their prey. The second, just as rapid, is to spit out the sea slug. After that, they never touch them again. The sea slugs must have left a very bad memory for their "taste buds." The sea slugs even protect themselves against bites: as soon as they are disturbed, they surround themselves with a white cloud of toxins that they release suddenly. Untouchable, repulsive, the sea slugs can therefore go on their way free of predatory attacks.

We have here another useful characteristic for evaluating the risks of introducing these mollusks: the sea slugs, indeed, depend on a specific food. Without the highly specific toxins of *Caulerpa*, they cannot survive.

That is not all. The most extraordinary feature of our sea slugs is their "metamorphosis" into plants! This entire process derives from the fact that, in tropical seas, they encounter *Caulerpa* infrequently. Having consumed a clump of *Caulerpa* algae, its only food, a sea slug must travel a great distance for several days without eating anything before it finds another clump of *Caulerpa*. Our little sea slugs survive these long famines by astonishing means. When swallowing the cytoplasm of *Caulerpa,* they ingest the chloroplasts of the algae. The chloroplasts are tiny organelles containing chlorophyll; they are found in the cells of all green plants.

These organelles are little factories that use solar energy, water, and carbon dioxide to synthesize sugar reserves necessary for plant life, all the while releasing a by-product of the chemical reaction: oxygen. Normally, all animals that eat green plants simply digest the chloroplasts contained in the plant cells. This is what we do when we eat lettuce, the cells of which are full of chloroplasts. But instead of digesting the chloroplasts, *Elysia* saves them and sequesters them in intestinal diverticula near the surface of the skin. After all the *Caulerpa* in a clump have been eaten, the sea slugs, now green, take off in search of a new clump, activating their solar batteries, represented by the chloroplasts that have been spared digestion and stored. From a classic vegetarian diet, the sea slugs switch to the plant mode of nutrition, spreading out their skin as best they can to capture the sun's rays. Their skin functions like the leaves of plants! If they are well sunned (daylight suffices), they are able to survive and crawl along the sea floor for more than a month without eating. But without light and without eating *Caulerpa,* they die in several days.

It would be as if after eating two salads we have become green and have the ability to subsist without eating for a month so long as we can sunbathe every day.

In fact, this extraordinary association between species has also arisen in several unicellular animals (the ciliates)[7] and, in more rudimentary fashion, in other organisms. For example, all lichens, attached to tree bark or forming crusts on rocks, are composed of two associated species: a fungus and an alga or plant-like bacterium. Similarly, many marine animals are associated with unicellular species containing chlorophyll. For instance, the coral reefs of tropical seas consist of colonies of invertebrate animals that would be unable to live without minuscule algae intimately associated with their soft flesh. If the temperature rises too much, as has happened the last few years in many tropical areas, the algal partner dies or leaves the coral. The departure of the plant cells that assure photosynthesis and thereby bring sugars to the coral constitutes a death warrant for the animal: the coral whitens, then disintegrates. These partnerships (alga–fungus, plant-like bacteria–fungus, and alga–coral) are associations in which one partner cannot survive without the other. We call such an association a symbiosis (from the Greek *sunbiôsi,* derived from *sun,* meaning together, and *bioum,* meaning to live). Our sea slugs have "invented" a more sophisticated symbiosis. Instead of cohabiting with tiny algae, they have selected and retained the only useful part of the ingested algae: the chloroplast. This is an unusual symbiosis, because one of the partners (the chloroplast) is not an individual (like a species of alga, for example) but an element of a plant cell. It is also an unusual symbiosis, because the partners are not found side-by-side but with one (the chloroplast) inside the cells of the other (the sea slug). The internal partner (the chloroplast) is used by the host (the sea slug cell). This is an *endosymbiosis.*[8] Symbiosis is a life in common between two organisms that ends by their becoming interdependent: one cannot live without the other and vice versa. Endosymbiosis is the most advanced form of symbiosis: one of the two partners lives *inside* the other (thus the prefix *endo,* from the Greek *endon,* meaning inside).

It is this phenomenon that is the most determinative step in the diversification of life on Earth. This step is very different from the gradual evolution of a species described so well by Darwin. It does not result from a succession of mutations with infinitesimal effects or from subtle interbreeding that allows a lineage of bacteria to eat better, to reproduce more, to proliferate, to dominate, to become a plant or an animal. This step is another departure in the evolution of life. It is a revolution, a schism, a new genesis.

To underscore for the class the conclusion of this story, I leave the blackboard and walk among the students. I examine some notes, which the students always prefer to hide, and I repeat:

"A revolution! A schism! Mark my words!"

Then I begin to tell the extraordinary tale of the animal and plant revolution. The first notion, which I emphasize, concerns the considerable differences between a bacterial cell and an animal or plant cell. This is a classic chapter in the program of instruction. In sum, here is what one must know about this subject.

Bacterial Cells and Animal and Plant Cells

All bacteria (archaea and eubacteria) consist of very simple cells. Inside these cells, there are no organelles; the contents of a bacterial cell are pasty. In most bacteria, there is only a single strand of DNA mostly rolled into a ball, located in the center of the cell. This is why all bacteria are known as *prokaryotes,* which means organisms prior to those that have a nucleus. *Karyon,* or, nucleus, refers to a "package" found in the interior of more elaborate cells (those of animals and plants) that contains the genetic information concentrated in chromosomes (this prefix is used in many words, such as karyology, the study of nuclei, and prenatal karyological examination, which detects chromosomal anomalies).

Species of flora and fauna are constructed of much more complex cells. They form the group *eukaryotes,* which means that their cells contain a true nucleus (the prefix *eu* means "well" or "good" in Greek). Algae, flowering plants, fungi, yeasts, amoebas, jellyfish, sponges, sea urchins, fishes, birds, and all of us are composed of eukaryotic cells, each with a nucleus. In the nucleus, a sort of porous box, chromosomes are found. Each chromosome consists of small units carrying information: genes. The genetic information is carried by DNA. The terms used by older biologists to describe this fundamental difference likened the cell to a fruit. Cells of bacteria (prokaryotes) had no pit, those of animals and plants (eukaryotes) did have one. This metaphor was apt.

If the nucleus contains the majority of a cell's DNA, playing the role of microprocessor, memory, motherboard, and operating system, other cards are inserted into the computer of life, the plant or animal cell. In effect, the eukaryotic cell contains other "packages" than the nucleus—the *organelles,* which have precise functions like that of a solar battery for the packages containing chlorophyll (that is, the chloroplasts) or that of a respirator for other packages (called mitochondria)—like independent cards added to the motherboard.

Mitochondria! There is a word that only biologists know. It is thanks to these organelles that all species of animals and plants can breathe in an oxygenated atmosphere.[9] These are the lungs of our cells. I propose to cosmetic firms that they popularize this term that deserves to be better known: a hair spray with mitochondria would suggest that the hair will be well aerated, just as a lotion with mitochondrial extracts would make the skin breathe better.

Once I have sketched out the two systems of life, I act as if I am reflecting on it. The students stop taking notes and raise their eyes. A course is a piece of theater. It is important to have a text in mind and to express it well. Moreover, it is necessary to convince the audience, to stimulate their memories. For that, it is necessary to empathize with the spectators, to put yourself at their level, to foresee doubts about a line of reasoning or difficulty in understanding it, to arouse their curiosity, to surprise them, to break the rhythm, to make them pay attention. The more unpredictable a lecture is, the more attentive they will be. When total silence reigns, I begin again in a solemn voice:

"The question to ask is, how can we explain the passage from one system to the other? How did bacteria become animal or plant cells? In other words, how did prokaryotes become eukaryotes?"

This is like trying to explain the brutal transformation of a pocket calculator into a Cray XD1 supercomputer. Researchers have certainly imagined a slow and progressive transition between the status of a bacterium and that of an animal or plant cell. Just a few unconvincing arguments, essentially genetic in nature, are the basis of this hypothesis of successive, progressive, gradual mutations from the ancestors of all cellular life (the bacteria) to a representative of the eukaryotes.[10] But this is a false lead!

The surprise is evident among some students, used to a well-structured course that follows a logical train of ideas and is nicely compartmentalized into numbered chapters. My presentation with apparently independent parts seems to disconcert them. Some must have been thinking, "After the big deal about the difference between plants and animals, his long tirade about the sea slug, and his emphasis on revolution, what's he going to hit us with next?"

So it is time to stop the questions.

I justify everything I have been saying: "In fact, to explain the revolution in the history of life, we can refer to the phenomenon of endosymbiosis adopted by the vampire sea slugs of the killer alga."

The Union of Bacteria

The metamorphosis of bacteria into animal or plant cells by endosymbiosis has been imagined from the beginning of the twentieth century, based on very meager clues.[11] It has become known as the theory of endosymbiosis.[12] This theory was reinforced and popularized beginning in the 1970s by Lynn Margulis, an American researcher who became famous thanks to her many publications and to the evidence she amassed using pictures taken with electron microscopes and data from genetic analyses.[13] I will summarize the stages in this phase of the emergence of new evolutionary pathways of life on Earth.

First, recall that the initial actors in the "revolution/metamorphosis" were bacteria, the first organisms to have colonized the Earth. Because of current knowledge on the behavior and nutritional modes of these simple organisms and on the environmental factors that reigned more than 2.7 billion years ago, the following two scenarios have been proposed.

How Mitochondria Arose

The first scenario, sketched by the founders of the theory of endosymbiosis, is set in the muck of a shallow lagoon (figs. 6 and 7). Bacteria are very abundant there, so abundant that, when they die, the remains of their organic matter accumulate at the bottom of the lagoon. Some opportunistic bacteria specialized in exploiting inert organic matter from the excretions of bacteria or the decomposition of their cadavers. They dissolved then assimilated these constituent substances of life. The bacteria that "ate" the organic matter did so by simple contact, just as fungi do. These bacteria stuck to organic matter and their digestive fluids moved both ways through their walls.

To improve the yield, it is enough to augment the contact area between the bacteria and the organic matter. For that to happen, an important transformation must have occurred. A bacterial species must have lost its usual carapace, a trait of the great majority of bacteria, to become supple, like modeling clay.[14] This first step allowed these bacteria to spread themselves out and absorb more organic matter by simple contact.

Having profited from this advantage, this fungal bacterium (a term explained in the preceding chapter) next acquired, by natural selection, the ability to deform itself to increase the contact area even more. A system of muscular microfibers[15] must have evolved to coordinate movements allowing the cell to crawl (once this individual digested everything under it, it moved to seek food not too far away).

This cellular behavior is not a fantasy. Nowadays, certain fungi use exactly this mode of nutrition.[16] Representatives of this group are soft and move by crawling over dead wood. Organic matter is digested by contact and energy-bearing molecules are transferred through the supple fungal membrane.

Subsequently, by dint of living and feeding in bacterial cemeteries, representatives of the soft, crawling bacteria tried another, more effective mode of nutrition. Continuing evolution led the supple wall of the fungal bacterium to incorporate bacterial cadavers.[17] The inert prey, therefore, found themselves intact, wholly inside the bacterium that had become a carrion eater. Chemical secretions allowed the digestion and assimilation

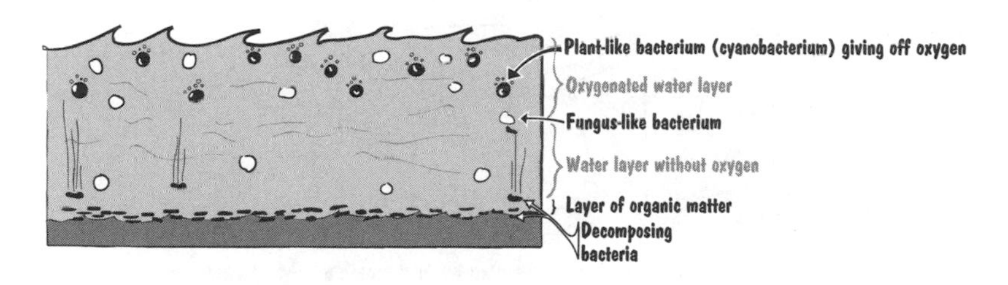

Some 2.7 billion years ago somewhere in the ocean near a coast...

A fungus-like bacterium gets ready to digest organic matter on contact

Several million years later...it then crawls in order to feed better

Several million years later...it now swallows organic matter (phagocytosis)

Figure 6. Endosymbiosis: the origin of the first animal cell. (Drawings by Alexandre Meinesz, arranged by Marjorie Meinesz.)

of the organic matter thus taken in. This system was very effective, because the contact area with the incorporated matter was greater than when food was simply covered. Thus, the fungal bacteria, having specialized in the capture of inert organic matter, became carnivores. They encircled, incorporated, and digested living prey, other small bacteria. This fundamental stage in behavioral evolution corresponds to a mode of life that

Figure 7. Endosymbiosis: the origin of the first animal cell (continued). (Drawings by Alexandre Meinesz, arranged by Marjorie Meinesz.)

is currently still common among certain animal cells, like amoebas and white blood cells.

The last stage is endosymbiosis. It arose when some living bacteria that had been ingested, instead of being digested, were spared by the digestive juices. Three explanations have been proposed:

- The carnivorous bacterium benefited from its domesticated prey and treated it as a sort of domestic slave.
- Certain prey, better constructed to resist the digestive juices of the carnivorous bacterium, took advantage of the latter by parasitizing it.
- Host and prey both benefited from living together (true symbiosis).

Of course, the bacteria did not "feel" anything and did not calculate the advantages and disadvantages of the association. Rather, this evolution occurred, as always, by a cascade of chance mutations, selected by environmental constraints that sculpted the lives and designed the behaviors of the two bacterial types.

The result is simple and identical, whichever hypothesis is correct: certain living prey became definitively acclimated to the interior of bacterial carnivores. The two cells—the host cell and the prey cell—divided systematically and indefinitely in lock step with one another so that the descendants of the host cells always contained "tamed" prey cells.

This is how the large bacterium (the host), a new type, accustomed to living in an environment without oxygen, was able to associate itself with a small bacterium endowed with physiological functions that allowed it to move in an oxygenated environment.[18] It is as if the host fungal bacterium, evolving solely in putrid water without oxygen, grafted on a lung to live closer to the surface near the light where oxygen was beginning to appear, thanks to the functioning of plant-like bacteria. Once surrounded and fully domesticated, the small, ingested bacterium lost its carapace. It became an organelle of the cell—the mitochondrion. Because of this act, the host beneficiary itself became a prototypical animal cell finally able to graze in the pastures of plant-like bacteria near the surface, in oxygenated water.

The second scenario was suggested in 1998.[19] It stipulates first that the emergence of the first animal cells occurred in water, near plumes of hot, sulfurous mud rising from the depths of the Earth or at the periphery of humid, warm interstices of the Earth's crust (fig. 8). There swarmed (and still swarm) bacteria using gas rising from the decomposition of rocks by hot water under pressure.

According to this hypothesis, an association for the joint use of energy resources and waste spurred the origin of symbiosis. The two partners were:

- A chemobacterium (archaebacterium). A "rock-eater," it nourished itself with hydrogen and produced organic wastes.

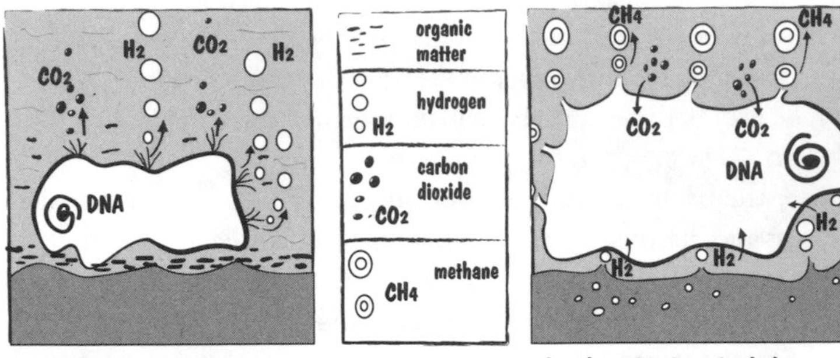

Figure 8. Formation of first animal cell: the hypothesis of the waste recycler. (Drawing by Alexandre Meinesz, arranged by Marjorie Meinesz.)

- A fungus-like bacterium (eubacterium). Smaller than its partner, it played the role of minority shareholder in assimilating the organic matter produced by the chemobacterium (its wastes) through its cell wall. Ubiquitous, it can live in water without oxygen, and, in that circumstance, expel among its waste products hydrogen, which is useful to its partner. It can also live in an oxygenated environment and, in this case, use oxygen to make by-products of energy production.

These two types of bacteria still exist today.

The two bacterial partners would at first have lived attached to one another, each feeding on the waste products of its neighbor (hydrogen for one, organic matter for the other). This association for recycling wastes led the "hydrogenivorous" bacterium to embrace the hydrogen-producer more tightly. In the end, it surrounded its partner.

The minority partner, the fungus-like bacterium, assured a direct supply of hydrogen to its host while taking its fuel (organic matter) inside the host cell.[20] This endosymbiosis produced a more rational exploitation of joint nutritive resources.

The host was increasingly enslaved by the faithful furnisher of hydrogen and increasingly avoided its normal feeding habitats (rocks decomposed by water and high pressure, releasing hydrogen). To ensure effective functioning of the partner living inside its body (its cell), the host changed habitat and food, as described in the above scenario. It consumed organic matter that the fungus-like bacterium sorely needed. The host became soft, able to assimilate organic matter by contact. Then it became able to ingest living, inert prey—it became a predatory animal cell.[21] In this new context, the host ventured toward the surface, into oxygenated water. The versatile bacterium, acclimated to its habitat inside the host, having furnished hydrogen so far, changed diet and waste products. It began to use oxygen and to excrete substances that served as energy sources for the host. This is how the captive bacterium allowed the host to breathe, to assimilate oxygen—the captive became a mitochondrion. And the two united bacterial partners then constituted an animal cell.

Whichever scenario for the endosymbiosis is correct, it is important to be clear that our own cells are ex-carnivorous bacteria. Inside them, we still find vestiges of this association and their feeding, because inside our cells swarm mitochondria, ex-domesticated bacteria that still reproduce. These bacteria exist in all animal and plant cells in oxygenated habitats. Without them, our cells could not live, so we could not survive. They are the lungs of our cells.[22]

These two simplified scenarios of endosymbiosis rest on much evidence:

- Mitochondria are the same size as bacteria that function similarly.
- Mitochondria have retained a certain degree of independence, for they divide like bacteria inside their host animal or plant cells.
- The experimental extraction of mitochondria from a cell leads to the death of the cell, which is unable to reconstruct these independent elements that have become indispensable.
- Nowadays, in all mitochondria of plant and animal cells, there is still a small amount of DNA of the bacterial type. But there is much less of it than in similar bacteria that have remained independent. This fact suggests that there has been a transfer of genes from the bacteria-mitochondria to the genome of the host cell nucleus (which has increased the number of genes of the latter). This hypothesis has been verified and is widely accepted by researchers.

How Chloroplasts Arose

This entire process must have begun nearly 2.7 billion years ago. At this remote time in the history of life, bacteria and prototypes of animal cells behaving as carnivores began to live together, at the interface of oxygenated and deoxygenated water. The second phase in the diversification of life began then with the co-opting, by means of endosymbiosis, of plant-like bacteria, those containing chlorophyll.

This process could have happened in this way.

Thanks to mitochondria, animal prototypes began to evolve in the oxygenated water of shallow lagoons where plant-like bacteria thrived (fig. 9). These plant-like bacteria were relatively immobile, easy prey for these voracious hunters. By ingesting plant-like bacteria and conserving them whole (only the rigid bacterial wall was digested), the vegetarian animal cell was able to transform itself into a plant cell.

In effect, by co-opting a plant-like bacterium, the predator no longer had to bother hunting prey. The vegetarian animal cell host exposed itself to sunlight, allowing its solar battery to function—the latter was the domesticated ex-plant-like bacterium, which had become an organelle containing chlorophyll.

The domesticated plant-like bacteria thus became chloroplasts (packages of chlorophyll), and the cells that used them changed status. *The*

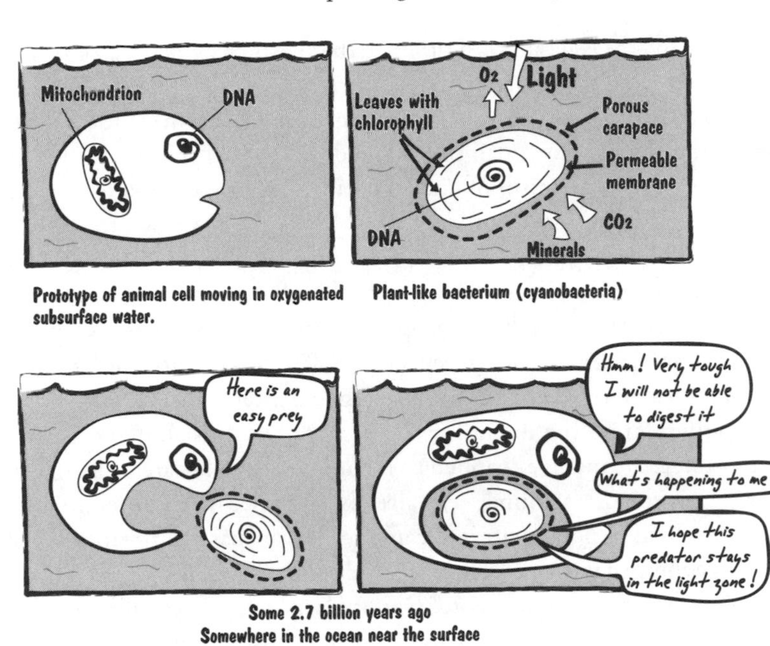

Figure 9. Origination of first plant cells by endosymbiosis. (Drawing by Alexandre Meinesz, arranged by Marjorie Meinesz.)

prototypes of herbivorous animals became prototypes of plant cells. To fix it in our minds, let us affirm that herbivorous animals are the basis of plants.

Every alga and green plant contains, in the heart of its cells, these ex-bacteria that give it its green color. Like mitochondria, chloroplasts contain DNA of a bacterial type. They replicate independently of their host cells in such a way that the descendants of the host cells always inherit them. Chloroplasts are the size of modern plant-like bacteria, and, if they are suppressed experimentally, the cell cannot synthesize them. Much evidence, some of it highly detailed, has accumulated that confirms this endosymbiotic process.

The bacterial origin of chloroplasts is even more evident in the few species in which the chloroplast is still surrounded by a solid bacterial wall.[23] In these species, the mechanism of endosymbiosis seems incomplete.

This is exactly the same process that allows the sea slugs to use the packages of chlorophyll aspirated from the alga. Without knowing it, the sea slugs have copied carnivorous animal cells: all they did was to remove, then put to their own use, the chloroplasts contained in the ingested fluids from the alga *Caulerpa*. The green chloroplasts, ex-plant-like bacteria, long domesticated by *Caulerpa,* have changed masters.

The only item on the menu for the first carnivorous animal cells was bacteria. Later, the menu expanded to include more elaborate cells. In this way, a prototype of an animal cell, already equipped with mitochondria and the resultant genetic material, came to devour passive prototypes of plant cells equipped with mitochondria and chloroplasts. For the first time, a eukaryote ate another eukaryote! Certain large prey were domesticated (everything is relative: in this context, cannibalistic animal cells could achieve diameters of several hundredths of a millimeter, while their prey were 10 times smaller).

A eukaryotic plant survived inside a eukaryotic animal, and this union persisted indefinitely among their descendants. These chimeric beings, therefore, result from two successive endosymbioses. Some lineages of present-day algae arose from this pair of endosymbioses. Among these, we find cells with two nuclei; the larger descends from the "cannibal," while the smaller, which has become inoperative and recessive, derives from the prey.[24]

The class is now absorbed and a bit fascinated by this natural history. All the students endeavor to take notes and copy the diagrams I have chalked on the blackboard.

"There are enigmas throughout this history," I continue.

"How did the nucleus come into being? How did the size of the genome multiply a hundredfold between bacteria and the prototypes of animal and plant cells?"

The Origin of the Nucleus

Today, if all biologists accept the endosymbiotic origin of mitochondria and chloroplasts, the hypothetical mechanisms for the origin of the nucleus are legion.

To understand the process of the multiplication of genes leading to the formation of a large chromosome or several chromosomes, geneticists have deciphered the genomes of very simple animal and plant cells. They were surprised to learn that these genomes are composed of a mélange of genes derived from several very different types of bacteria. How can this fact be explained? Two hypotheses are especially cogent:

- A predator was able to "adopt" certain genes of ingested bacteria. This ability is demonstrated by the endosymbioses of mitochondria and chloroplasts. This same process has occurred several times with smaller parts of genomes of various bacteria that were not domesticated but were simply swallowed.
- Viruses of bacteria[25] were able to hijack a small gene from a bacterium, transport it, and graft it into another bacterium or a prototypical eukaryotic cell. The possibility of such an event has also been proven. Viruses are degenerate bacterial parasites of other bacteria or of eukaryotes. They can no longer live and reproduce by themselves. Their lives and reproduction depend wholly on the other bacteria or animal or plant cells that they parasitize.

From transplant to transplant, the genome was thus augmented, forming a heterogeneous patchwork woven into a large chromosome that can be decomposed into distinct units.

But in plants and animals, the chromosomes are in a porous package, composed of a double membrane: the nucleus. Theoreticians explain its formation as resulting from an invagination of the cell membrane, which therefore forms a diverticulum enveloping the chromosome or chromosomes. Others have hypothesized that the nucleus was also the fruit of an endosymbiosis. The nucleus would, in this view, be an ingested bacterium (fig. 10). Its genome could have been enriched by that of its host, by that of mitochondria and other prey, and by those of viruses. These two hypotheses lead to the same result: the formation of a prototypical unicellular animal with a nucleus containing one or more chromosomes.

A bit to the right, towards the fifth row of the auditorium, a group of students is discussing something. One of them raises her hand.

I call on her: "A question?"

"And what about sexual reproduction? When did that arise?"

This is a good question. It is strange to note the prudence and discretion on this subject in the scientific literature. There is not much evidence, so imagination has been given free reign.

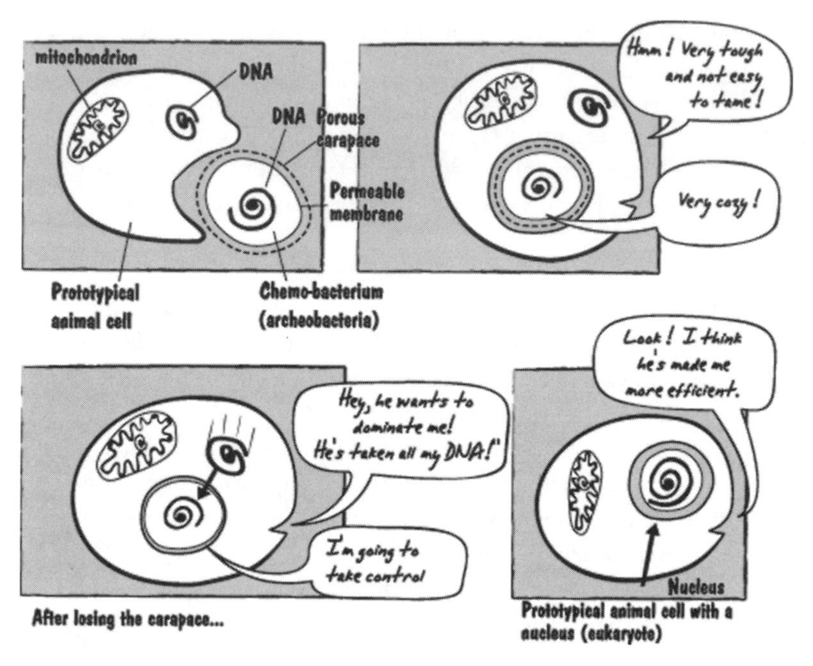

Figure 10. Origination of the nucleus by endosymbiosis. (Drawing by Alexandre Meinesz, arranged by Marjorie Meinesz.)

The Origin of Sexual Reproduction

We must remember cell division in all plants and animals occurs in two different modes. First, the simplest. When a cell divides into two, everything is duplicated unchanged (a cell with two chromosomes produces two cells, each with two chromosomes).[26] This mechanism operates in all our cells. This is what allows us to grow and to renew our cells with identical replacements. This is also the process common to all bacteria, where genetic information is duplicated unchanged when the cell divides. This is a form of "vegetative," asexual reproduction.

Another, more complex type of cell division leads to a halving of the number of chromosomes. This mode, with reduction of the genome, is the prelude to the sexual reproduction found in all plants and animals. This *reduction* division occurs in the reproductive organs (in humans, in the testes and ovaries). Beginning with a cell containing two similar sets of chromosomes, the process ends with cells containing a single set. For example, our own cells contain 46 chromosomes, composed of 23 pairs of homologous chromosomes. Their division, according to this mode, produces cells containing 23 chromosomes each. These cells are our spermatozoa and eggs. It is a "divorce," the halving of the genome, the equitable reduction in the number of chromosomes.[27]

Now, we know that our reproductive cells (spermatozoa and eggs) can survive only if they encounter one another. They then unite their reduced numbers of chromosomes to reconstitute a single cell containing two sets of homologous chromosomes. This is a "marriage,"[28] a union, the ultimate step in sexual reproduction.

To explain the origin of this double process (chromosomal divorce and remarriage), scientists say we must imagine, speculate, when definitive proof is lacking.

Everything must have begun by a doubling of the number of chromosomes.

The passage from a simple set of chromosomes to their doubling can be explained by an error in the timing of their copying.

Cells of all organisms all have the ability to divide in two; in this mechanism, all genetic information should be preserved and transmitted. To this end, a process—termed *mitosis*—allows the duplication (multiplication of their number by a factor of two) of the chromosomes just before the whole cell divides. The genetic information is thus transmitted unmodified and equally shared between the two daughter cells.

But we can imagine a poor synchronization of the division of a cell and the division of its chromosomes. If the chromosomes double in the cell but the cell does not divide, the game is won! We have passed from a single to a double set of chromosomes in the nucleus.[29]

But there is another hypothesis, based on the cannibalistic behavior of the precursors of animal and plant cells.[30] Imagine that two identical, independent animal cells meet; they are voracious prototypes from the same lineage, each with a nucleus containing just one set of chromosomes (fig. 11).

They are ready to eat one another. The hypothesis stipulates that, instead of mutual devouring, the two cells fuse.

This behavior (an abortive cannibalism) leads to a sort of amalgamated symbiosis, a happy union. Subsequently, in the single cell resulting from the "marriage," the two similar nuclei approach one another and also fuse.[31] There are therefore two identical sets of chromosomes mixed together without reduction in their number or elimination of any of them: this is the doubling of chromosome number.

A cell equipped with two sets of chromosomes is more vigorous, more effective than the original cells, because the role of a gene that might be

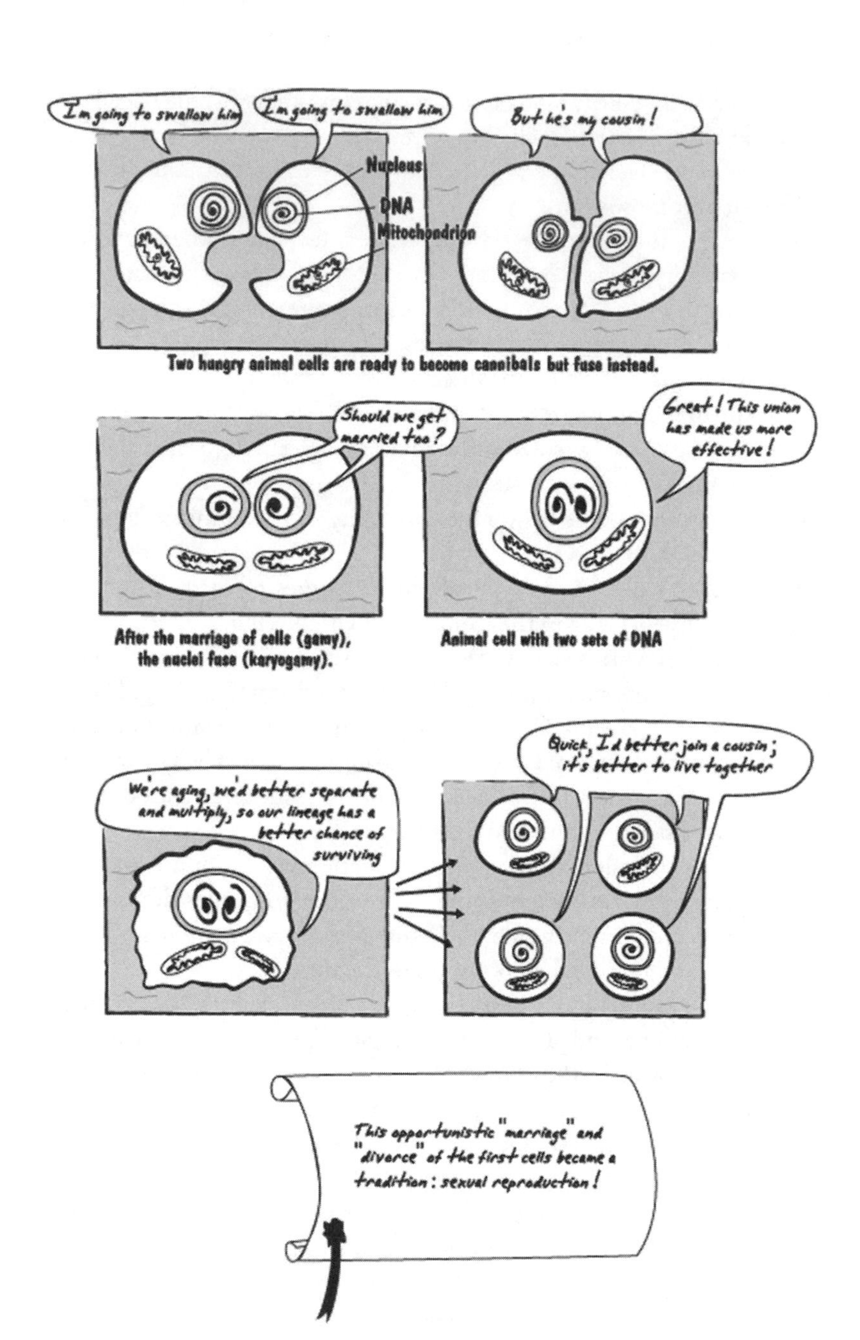

Figure 11. "Marriage" and "divorce" of cells: first appearance of sexual reproduction. (Drawing by Alexandre Meinesz, arranged by Marjorie Meinesz.)

deficient on one chromosome can be compensated for by the information carried on its homologous chromosome. The cell will survive and dominate otherwise similar cells that have only one set of chromosomes.

But the vigorous cells retain the potential to reduce the number of chromosomes. Therefore, during a period of stress or a habitat change, the cell containing two sets of chromosomes divides by two (cell, nucleus, and chromosomes), which leads to the formation of identical cells carrying a single set of chromosomes (this process is called *meiosis* or *reduction division*). From this point on, these cells should be called sex cells, because they have only one destiny and route to salvation: to find an identical, compatible cell to marry. In fact, to begin anew the scenario of the abortive cannibalism of their direct ancestors. This is how sexual reproduction arose.

Sexual behavior appeared in the first independent animal and plant cells. At the outset, it was only a sort of atavistic behavior, a memory of a solution to the problem of living differently, in a simple but too fragile state, because the cell had only a single set of genetic information.

As for me, I greatly favor this hypothesis explaining the universal division of animal and plant cells. All their sex cells behave similarly: they have soft walls, they come together, and they fuse (fig. 12). In humans, the large egg easily absorbs the small spermatozoon. Once the spermatozoon is incorporated in the egg, its nucleus sticks to and then fuses with that of the egg, producing a combined genome in a single nucleus.

This marriage between the two sex cells is really a sort of *fusional symbiosis:* the spermatozoon and the egg cannot survive unless they fuse.

To illustrate this hypothesis, I should refer to a remarkable behavior, very similar to what must have happened at the time of the original fusional symbioses. It can be seen in certain groups of fungi, especially those we eat.[32]

Ordinary mushrooms (with a cap and a stalk) do not have sex cells produced in specific organs, but cells of some of their underground filaments (the mycelium) *have only one nucleus, which contains a single set of chromosomes.*

The cells of these mycelia have the extraordinary capacity to fuse with similar cells that are compatible.[33] It is as if all the cells of these filaments carrying a nucleus with a reduced genome are identical sex cells. But in this case, the marriage of two cells is imperfect: even though two cells have come together and fused to form a single cell, the nuclei of the two fused cells *remain independent, side by side.*[34] The fusional symbiosis, the merging of the contents of the two cells, *does not lead to the union of the nuclei* (fig. 13). The two cells have "mated" to produce a single one, but the two nuclei inside the new cell have "separate bedrooms."

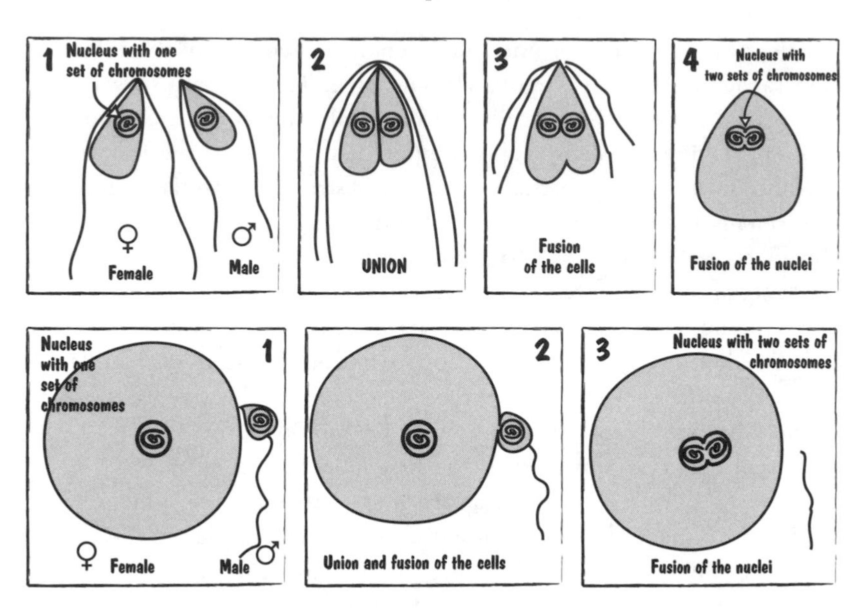

Figure 12. Present-day sexual reproduction of multicellular organisms with union and fusion of membranes (plasmogamy) and nuclei (karyogamy). *Top*, In some species of green algae, the cells involved in sexual reproduction (gametes) are identical. *Bottom*, In all multicellular animal species (metazoa), the immobile female gamete (ovule) is much larger than the highly mobile male gamete (spermatozoon or sperm cell). (Drawing by Alexandre Meinesz, arranged by Marjorie Meinesz.)

The cells containing the two nuclei develop more complex structural properties than the filament cells that have a single nucleus. Only the underground filaments with two nuclei can produce a large mushroom. The mushroom, as one can imagine (given its cap and stalk), is but a compact structure composed of mycelia with cells containing two nuclei; some of these cells are destined to produce spores. These spores are produced under the cap, on the lamellae (gills), or in the pores of the cap (for boletes). In this specific place, the cells containing two nuclei change and the two nuclei fuse just before spore formation. The sets of chromosomes then quickly mix and individual chromosomes exchange portions with their homologues. The newly constituted nucleus, which contains two sets of chromosomes, divides as the cell undergoes meiosis to form spores, each equipped with a single nucleus, once again carrying just one set of chromosomes. The spores germinate in the soil to produce mycelia, filaments of cells that each contain just one nucleus. This is the life cycle of ordinary mushrooms.

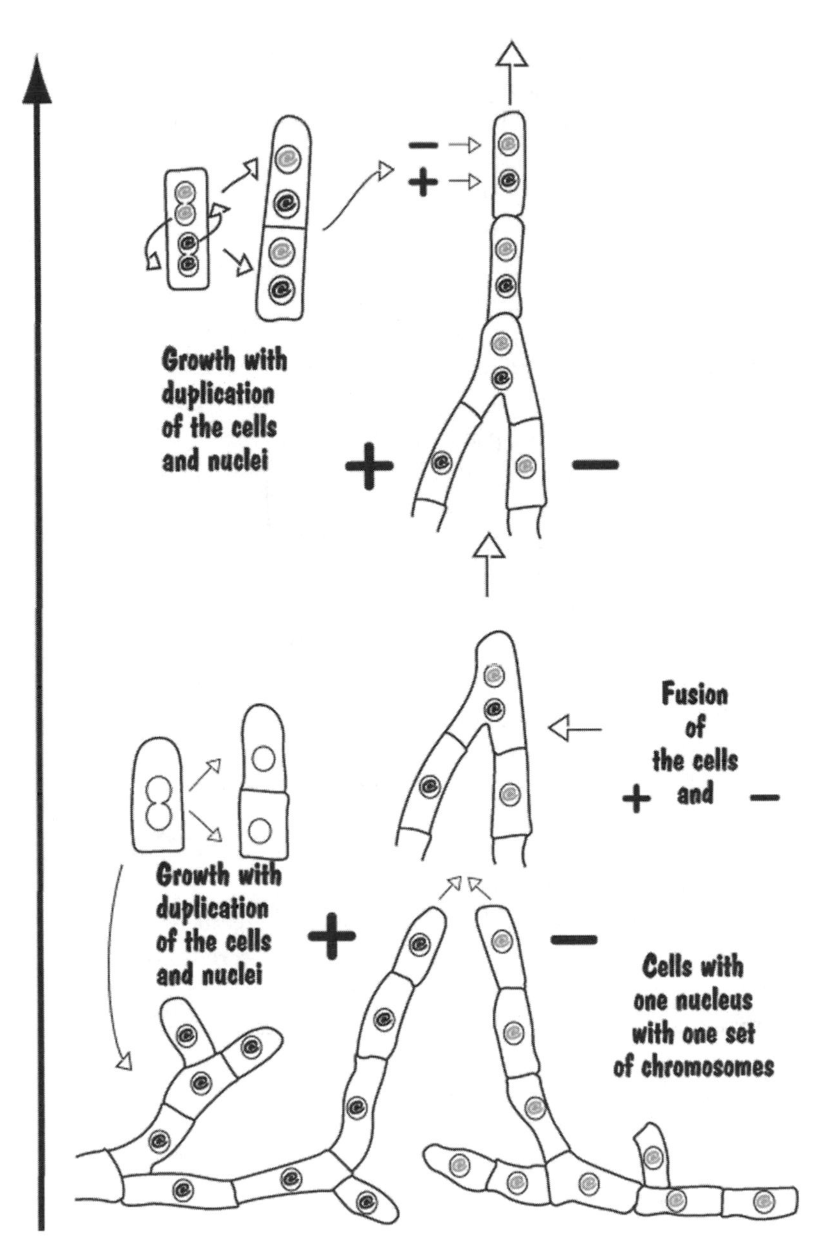

Figure 13. Present-day reproduction of eumycetes and basidiomycetes in which the "marriage" occurs in two stages. The filaments arising from different spores do not have sexual characteristics but are able to unite (or not unite). Compatible filaments (a + is compatible with a −, but not with another +) unite and fuse (plasmogamy or somatogamy), but the nuclei do not unite. Filaments composed of cells with two nuclei (dikaryons) make up the visible fungus (carpophore). Karyogamy, followed immediately by meiosis, occurs before the spore formation, which takes place in the lamellae under the mushroom cap. (Drawing by Alexandre Meinesz, arranged by Marjorie Meinesz.)

In several algal lineages, the male and female sex cells swim like spermatozoa, and they seem identical; only subtle differences distinguish male and female gametes: these differences are the first sex characters (in the majority of algae, the sex cells of the two types are very different—the male cells are small and have flagella, while the female cells are larger and often lack flagella). In all these cases, for two sex cells to unite, they have to be of the same species, containing the same types of chromosomes.

The most extraordinary case of atavism that I have seen is in algae dear to my sea slugs: those of the order Caulerpales. These green algae look like ordinary plants, but they do not have cells—there are no internal cell walls. They are just a tube shaped like a plant. In the tube, there is a sap (cytoplasm). Mixed in this sap are nuclei, chloroplasts, mitochondria, and all the other components of plant cells.

When the time comes for sexual reproduction, clear the decks! It is a marriage ceremony.

The nuclei, each containing a single chromosome set, assemble all the constituents of the original cell: chloroplasts and mitochondria. Each nucleus then surrounds itself and its acolytes with a membrane, thus forming a sex cell. The entire contents of the alga are thereby transformed into reproductive cells. At a signal (daybreak), they escape to the outside—open water. The alga, emptied of its contents, no longer exists: it is now only the inert, bleached wall of the empty tube, which quickly disintegrates.

To survive in the water, the minuscule sex cells have only one hope: to find a similar and compatible sex cell with which to fuse, to marry. The union will lead to the formation of a new alga.

This is the only group of large organisms where all the nuclei of an individual participate in the traditional marriage ceremony. When the individual comes to the end of its life, or when environmental conditions deteriorate, it scuttles itself to perpetuate its genes. Its whole contents are transformed into eggs or sperm! The entire living substance is fragmented but continues to exist; only its "skin" is abandoned.

The processes of reduction of the number of chromosomes and then of their fusion allow exchanges of chromosomes and other parts of the mixture; similar whole portions can pass from one chromosome to its homologue when they line up or abut one another (this mechanism of recombination is known as *crossing over*).[35] This whole process leads to a genetic mixing. Animals and plants have therefore acquired a mechanism that stabilizes certain mutations, erasing the most aberrant ones while permitting the fixation of traits that differ from individual to individual. This system, favorable to innovation, was destined to become established and used. The great mixing of genetic information is advantageous, allowing evolution in a changing, hostile environment. Tough natural constraints have led to the selection of species with this fantastic potential for variation. Sexual reproduction was thereby imposed on life and became obligatory (in most species) in the renewal cycles of animal and plant species.

Only in unicellular animals and plants living in stable environments is there little recourse to sexual reproduction. In such species, cells usually divide just as they do in bacteria, by simple duplication of the cells and their genetic material.

The end of the lecture approaches; I sense it without glancing at my watch. Attention wanes. I have to reconnect with my audience. I stand still for a moment. Until the students, surprised, all stare at me.

"It's not finished!"

I encourage them:

"Look. Only five more minutes and the entire history of endosymbiosis will be over. In fact, there is still one more endosymbiosis to cover."

The Origin of the Flagellum

In order to hunt, the bacterium, prototype of the animal cell, had to move. First it had to crawl. It would then be able to float in the water but not to orient itself. It is very likely that another endosymbiosis allowed the acquisition of a trait characteristic of most animal and plant lineages: the flagella or cilia that permit the cell to move.[36]

Some present-day bacteria are mobile. These are shaped like strips with contractile properties that allow propulsion.[37] This ability probably came from a similar bacterium that was ingested. A part of this bacterium, remaining outside, was used by the host bacterium to move itself. There is not yet enough evidence supporting this hypothesis of endosymbiosis for it to be fully accepted, but it is noteworthy that all cilia and all flagella of distinct lineages of animal and plant cells have the same very complex internal structure and the same functional mechanism.

Some lineages (genealogical trees of life) completely lack cilia and flagella, notably, red algae and some fungi (including mushrooms; the ones we eat are part of this group).[38] In these groups, mobile bacteria were never part of the menu of the carnivorous bacteria that were their ancestors.

In the green plant lineage, we observe that there are spermatozoa with a flagellum in the green algae, the mosses, and the ferns (they have to swim in water to reach the egg, which is why there are neither mosses nor ferns in dry places). In flowering plants, which arose later than algae, mosses, and ferns, the pollen grain (comprising several cells) produces the male gametic cell. The pollen grain is carried by the wind or insects to the female flower organ, the pistil. The male gamete forms in the pollen grain, and it joins the egg thanks to the germination of a tube (the pollen tube) that bores into the pistil to touch the ovule, the structure that develops into a seed. As there is no need for flagella, they have been suppressed in the course of evolution.

The Second Genesis of Life

Let me sum up the situation: The ancestors of animal and plant cells were bacteria that became associated with smaller bacteria (destined to become mitochondria, which became, as it were, the lungs of the cell). These soft-bodied ancestors became carnivores. By gene transfer and an increase in the number of genes (through fusion with their sister cells), they were endowed with pairs of chromosomes concentrated in a nucleus. They became unicellular animals that reproduced sexually. Among these, certain ones fed on diverse plant-like bacteria that ended by being included in the carnivorous cells as chloroplasts. This series of associations gave rise to various lineages of chlorophyll-containing plants. Other of these carnivorous bacteria, which were more voracious, became cannibals and ingested and tamed the prototypical animals and plants, partly domesticating them. These gave rise to other animal and plant lineages. Among all these lineages, certain ones probably co-opted mobile, strip-shaped bacteria that became their means of propulsion (cilia and flagella).

This summary emphasizes that the emergence of our flora and fauna was a consequence of unions: they are only a grand mélange of various bacteria.

Once again, I put down my chalk. I stress this conclusion.

"We are here thanks to successive endosymbioses. Animal cells, plant cells, our cells—all our cells are ex-bacteria that still contain within themselves other, domesticated bacteria!"

Whispering in the lecture hall. Astonishment of students who have grasped the application of the theory of symbiosis to our own cells. Others get a laugh out of treating their neighbors as a swarming pile of bacteria.

The diversification of the fauna and flora is therefore the result of novel assemblages formed by chance by a lineage of cells that had invented the process whereby each cell united its life with the life of another cell. The carnivorous and cannibalistic bacteria appeared more than 2.7 billion years ago, and they no longer exist in our marshes and oceans. However, today they would have lots to eat, because their basic food supply (all bacteria and microscopic algae) has never dried up. This group of creative artists has disappeared forever, leaving only a small part of their work, perpetuated by distant descendants that make up the whole of our present fauna and flora.

In this entire process, small mutations and natural selection have continually intervened. Bacteria whose genomes conferred the ability to swallow prey therefore had an advantage in the habitat they occupied, in comparison to those that simply absorbed organic matter by contact. Among the bacterial carnivores, those that preserved and used their prey without digesting them benefited in one fell swoop from a great advantage over those who fully digested everything they ate.

Multiple combinations of bacteria must have appeared and confronted both environmental conditions and competition with one another. Myriad microscopic hybrid or chimeric beings, prototypes of animal and plant cells whose form and functioning we will never know, must have constituted, along with bacteria, the biodiversity of this epoch.[39] Associations poorly adapted to the changing environmental conditions did not survive. The viable associations had an ephemeral life, being transformed by evolution into other unicellular animals and plants. Thus the endosymbiotic associations that constitute modern animal and plant lineages make up an infinitesimal part of the associations that saw the light of day in the distant past and whose growth was interrupted in a changing, hostile world.

Today, we do not distinguish more than several dozen independent animal and plant lineages that arose during this epoch, long past in the history of life. They arose from many different and successive combinations and have survived cataclysms,[40] competition, and the contingencies of evolution. Each is characterized simply by a different life force. The depths of their cells do not contain exactly the same ingredients. They are the result of different menus, tested by the original carnivorous bacteria. This is why we must admit, for example, that a green micro-alga is closer to a palm tree or to a fern than to a minuscule brown alga. The engine of life, the cell, functions identically for the green alga, the fern, and the palm—these are representatives of the same evolutionary lineage; their cells are fundamentally different from those of a brown alga (this type of alga is often used as a background for the presentation of seafood in restaurants[41]). The two lineages—plants and brown algae—arose from different combinations; the bacteria ingested nearly two billion years ago by the prototypes of these two lineages were not identical. Certain endosymbiotic combinations have produced lineages that are half plant (they have chlorophyll) and half animal (they nourish themselves only by ingesting organic matter).[42]

It is for all these reasons that it is impossible to give a precise definition to fauna and flora, words created by man thousands of years ago to distinguish plants from animals. This informal distinction simply does not adequately capture the diversity of living systems composed of cells with nuclei, visible to both the naked eye and the optical microscope.

Here, the lecture ends. Three times I have erased my drawings and technical terms that covered the two mobile blackboards. My hands are covered with multicolored dust, traces of colored chalk. My jacket is sprinkled with chalk powder. It is time to finish.

"OK! You now understand why the process of symbiosis marks a schism in the evolution of life on Earth. Today, the flora, fauna, and man all retain in their cells the descendants, indefinitely reproduced, of bacterial feasts. In our cells, which are ex-carnivorous bacteria, domesticated bacteria that were co-opted several billion years ago still swarm. One of the major behaviors of animals and plants—sexuality—is atavistic; it is basically a cannibalistic behavior transformed into one of union, and it dates from the first endosymbiosis.

"You understand now why I enjoyed teasing you with my questions about the differences between fauna and flora and why I told you about the vampire sea slug of the killer alga, an animal that 'reinvented' endosymbiosis[43]—there is the justification for the word 'revolution.'

"Yes! It was a revolution, because the appearance and development of the fauna and flora are not only the fruit of a slow evolution beginning with one bacterium or even several bacteria that represented the single common trunk (or few trunks) we think of as the tree of life. It was not only small genetic mutations selected by nature that produced the transformation of a bacterium into an animal or plant cell. Flora and fauna were not derived from slow individual modifications but from unions, successive *additions*. Additions born of parallel harmonious evolutionary paths: co-evolutions.[44] These friendly or hostile mergers have produced advantageous combinations. A bacterium A associating with another bacterium B produces a new individual C. This one ingests and domesticates a third bacterium D, yielding a new individual E.

$$\mathrm{A} + \mathrm{B} = \mathrm{C} \quad \text{and} \quad \mathrm{C} + \mathrm{D} = \mathrm{E}$$

"These unions have added together the abilities and potentials of the associates; they have also produced the power of the resultant cells.

"Sexual reproduction, a process of fusional endosymbiosis, is a new mechanism, a new motor of evolution. The union of cells allows for genetic mixing and blending in the mélange of chromosomes of both male and female origin. The doubling of chromosome number and the mixing of genetic information from the two united individuals (like an egg and a spermatozoon) confer on animal and plant cells both stability and creativity.

"All this is revolutionary, because it led to the emergence of flora and fauna."

The students stand up. I was secretly gratified to have been able to transmit such a beautiful natural history—a recently understood history that I had not learned when I was sitting in the same lecture hall at the University of Nice–Sophia Antipolis.

Everything that concerns the origin of bacteria on earth and the process of endosymbiosis that led to animal and plant cells constitutes the key to understanding the essence of life. These are the two major events that have led to the presence and diversification of life on Earth. These are two distinct geneses. Everything else is explained by a long series of transformations and catastrophes, the mechanisms of which, now much better known than when I was a student, are presented in the following chapters.

CHAPTER 5

Vermeer and Van Leeuwenhoek

Luck, random chance, risk, and contingencies in evolution

Johannes Vermeer, a painter widely admired today but poorly known in his own time, was born in 1632 in Delft. Delft was then a bustling little town in the United Provinces of the Netherlands, attached to the House of Orange-Nassau and a military bastion of the resistance to the Spanish invasion. The city was known for its art trade and its ceramic and tapestry workshops. In this environment Vermeer created a new style of Dutch painting. His depictions of the local bourgeoisie are strikingly realistic and attest to a refined sensitivity. Among the 33 paintings today attributed to him[1] are many representations of wealth and of artistic productions. Music is illustrated by young women depicted with various instruments.[2] In several paintings, we can admire carpets richly decorated with flowers posed on tables or hanging on walls. Several compositions in which women write or read letters show the art of writing, reserved for the elite. Paintings decorated the interiors of homes depicted by Vermeer, and one of his paintings has as a subject a painter in front of his easel.

But Vermeer was also sensitive to the art of the sciences: on six of his canvases, large maps are suspended on walls.[3] They express the sum total of geographical knowledge that was, by then, very advanced. Similarly, two paintings each portray a scientist: an astronomer in front of books, brushing his hand over a celestial globe (plate 6), and a geographer facing a naval map, holding a pair of dividers against a background of an armoire, books, and a terrestrial globe (plate 8a). Art historians agree that *The Astronomer* and *The Geographer* use the same model for the subject. He closely resembles an authenticated portrait of Antoni van Leeuwenhoek

Figure 14. Jan Verkolje, portraits of Antoni van Leeuwenhoek (1868). *Left*, Oil painting on canvas. Rijksmuseum, Amsterdam. *Right*, Verkolje reproduced the painting in a magnificent copper engraving, with one change: instead of the dividers, Van Leeuwenhoek holds one of his microscopes. A mezzotint print from this engraving is a mirror image of the painting (whereas in the painting, the dividers were in Van Leeuwenhoek's left hand, in the print, the microscope is in his right). (© Rijksmuseum, Amsterdam.)

painted by Jan Verkolje, in which the scientist also appears, dividers in hand, beside a terrestrial globe (fig. 14).

Van Leeuwenhoek, who was born in Delft the same year as Vermeer, was the executor of the painter's will.[4] The scientist was known for his passion for geometry, astronomy, and optics, and he became famous for a high-performance microscope that he built. With this instrument he was the first person to discover microbes and their astounding diversity.[5]

Although he had no scientific training, he was the first to observe and report that the color of our blood came from red globules, that semen has billions of "vermicules" that swim thanks to a whip (these were spermatozoa), that stagnant water contains myriad moving "animalcules" with cilia (infusoria), and that "globules" swarmed in the tartar that forms on our teeth (these are bacteria). All these microscopic organisms were subsequently named *monads*.[6] Despite his great uncertainty interpreting the microorganisms, we can only imagine how he marveled at what he saw.

In 1665, an English scientist, Robert Hooke (1635–1703), studied a thin section of cork using another version of a microscope. He observed that this inert plant material is composed of a mosaic of quadrangular boxes that he called *cells*. Van Leeuwenhoek confirmed this observation 12 years later. Subsequently, other researchers established that all living things are made of cells.[7]

In one fell swoop, a hidden scale of life was disclosed to Van Leeuwenhoek. For the first time, life became abstract, as its intimate structure was revealed and understood by means of an optical tool that was then very rare.

Nowadays, more than three centuries after Van Leeuwenhoek's observations, we can still discover and view remarkable cellular structures, thanks to the power of sophisticated optical or electron microscopes and with the help of advanced methods of lighting, fixing, staining, and developing preparations.

Van Leeuwenhoek, living in an environment in which all forms of artistic expression were valued, was especially sensitive to the shapes, colors, and diversity of microscopic life. The splendors of nature, the structures and colors of living organisms, animal and plant behaviors, odors and noises of life, the many shapes and expressions of the faces and bodies of women and men have always inspired writers, philosophers, musicians, and painters. For centuries, we have marveled at the beauty of nature, and, for ages—whatever our beliefs—we interpreted all living organisms as divine creations. This admiration for the richness of life was heightened in the Dutch provinces. Shortly after Vermeer's birth, the discovery of a flower, the tulip, drove many Batavians mad, exciting everyone from peasants to wealthy businessmen. However, it cost a fortune to own one; for one bulb, people sold their houses, went into ruin.

Painting indoor scenes, where art is magnified, and a superb view of Delft (highly appreciated by Proust in *À la recherche du temps perdu*), Vermeer created unique works, posed, static scenes that we still admire centuries later. At the same time, Van Leeuwenhoek discovered moving, variegated cells, multitudes of ephemeral, changing lives. He was the only one to see them, and he was also filled with wonder.

Van Leeuwenhoek, though an amateur scientist, was the first to observe the unicellular microorganisms that are the origin of life. These cells can live independently or can agglomerate to constitute organisms of large size. These are the bricks of living constructions, the elementary material of the architecture of visible plants and animals. The order in which cells are joined and their specialization for particular functions confer a consistency on organisms large enough to be visible. By the internal changing genetic information, shaped in the yoke of many physical and chemical constraints and tested by the external environment, the architecture of life is determined. The result is an extraordinary diversity of shapes, functions, and colors.

The Transformation of Species through Time and Evolution

In the first chapter, I summarized the surprising affiliations between apes and man and the wealth of knowledge on this subject. Today our animal ancestry appears logical and evident. Nevertheless, it still inspires debates and, here and there, incredulity.

The description of the transformation of prototypes of animal and plant cells into apes, whales, insects, flowering plants, or ferns is less surprising because it seems too abstract. "That must have happened that way, but I can't explain it," is the answer I get most often when I question friends who are not biologists.

However, it is a matter of the same processes that led to the changes between other apes and man. There were many transformations that, over millennia, produced the blossoming of life. But if the time it took for apes to evolve into man is estimated at several millions of years, the time that passed between the prototypical animal cell and the apes is estimated at 2.7 billion years. It is a question of another time scale and another scale of transformations.

Everything began with the appearance of animal and plant cells. Changes, transformations, and adaptations are terms with no reference to time that we often use to describe what happened next. *Evolution* introduced the notion of time; this word better captures the mechanisms that have taken so long to modify and diversify life.[8]

People who are not professional biologists often do not grasp the passion that the subject of evolution has inspired—and the mountain of books, articles, research, and controversies that it has created—since the writings of Lamarck and Darwin.[9] The craze for this subject is far from exhausted. Even today, monthly and quarterly journals are entirely devoted to this theme, and popular works proliferate. Natural historians, paleontologists, botanists, zoologists, embryologists, and, these last three decades, geneticists and biochemists continue to deepen our understanding of how species are transformed. These scientists belong to different disciplines and are often quite ignorant of one another's areas of expertise. They have added stones to the edifice of evolutionary knowledge in disparate fields, introducing subtle principles not easily understood except by close attention to linguistic nuances, new technical words, and even widely established metaphors. These students of past life have constructed new concepts[10] that integrate many advances significant to our detailed understanding of the process of evolution.

Luck, Randomness, and Contingencies

In their analyses, three words recur often; they express in different ways matters that are fortuitous: luck, randomness (or chance), and contingency.[11] Of course, luck, randomness, and contingencies can lead to bad outcomes as well as fortunate ones. These terms are often confounded in current usage. For those who describe evolutionary phenomena, they have different connotations.

The term *luck* should be reserved for unforeseeable events that are not expected in reference to some goal.[12] The farmer plowing the earth and finding a piece of gold is the beneficiary of good luck—the luck of having turned over the earth in exactly the place where the piece of gold was located, even though looking for it was the furthest thing from his mind. Of course, luck can also be bad; the farmer might lose his piece of gold gambling. Aspects of lucky events are predictable; even if the farmer did not know there was a piece of gold in his field, sooner or later he would have found it. We could calculate the probability of his finding it within a given amount of time by considering the area of the cultivated field, the frequency of plowing, and the size of the piece of gold.[13]

A random or chance event[14] is an unforeseeable event, but one that is expected given enough time. It concerns a known event, but no one can predict when it will occur. The term *random* often has a negative connotation—a house built too close to a riverbed risks destruction in the hundred-year or thousand-year flood (no one can say exactly when that random event will occur, only that on average one flood will occur during an interval of 100 or 1000 years). But a random happening can be either good or bad: a person who plays the lottery cannot predict the result of a particular draw, but in playing the game he or she is sure to have many losses and some wins (which was not the case with the farmer; he could not expect to find a piece of gold, ever). Games of chance are thus games of risk.

Luck and random events share the trait of being events that follow simple rules of probability. They apply above all to events, identified as such or not, that occur in probabilistic fashion. Paradoxically, one can say that in a specified time a lucky or random event will occur with a probability that can, in principle, be calculated.

It is important to understand the term *contingency*,[15] used especially in the plural, in the sense of an unforeseeable, indeterminate sequence of events in which each event is tied to another as cause and effect. This is a matter of events with incalculable probability—a random event for which no prediction is possible. We have no idea of the events that are going to follow one another, we do not know when or where each unknown event will occur. The ways they are juxtaposed are utterly matters of chance. The definition of the term *contingency*, which is not intuitive, can be illustrated by a simple fictional tale.[16] One morning, Bill missed his regular bus and took the next one. The same day, Julie missed the plane that was supposed to take her home. The next plane was leaving later that evening, so she decided to take a taxi to visit the city instead of waiting the entire time at the airport. But her taxi's electrical system short circuited and broke down, so Julie then took a bus, where she met Bill, and the two fell in love. In everyday terms, we say that the meeting, followed by the marriage of Bill and Julie, was the fruit of extraordinary luck. They are so well-suited to one another that we even doubt that it *was* luck. It seems as if their meeting was fated; many people who believe events are guided by a mystic destiny would go so far as to say it was directed by superior forces. For scientists, it was not blind luck at all; it was an illogical sequence of pure coincidences, totally unforeseeable. No mathematical law could have predicted the succession of independent events and the happy denouement (even without taking into account all the other events that could have happened to derail the sequence that occurred, but did not).

In nature, harmony is the result of contingent events. If an equilibrium is overthrown or destroyed, the original harmony will have an infinitesimal chance of being reconstituted identically. If it were possible to turn back the clock, to begin at the beginning, we would find a different succession of events and therefore a different final result. This holds true for our individual lives, for the lives of species, for the history of life.

Luck, randomness, and contingencies are three words—not always well understood and properly used—that figure in all the main conclusions of a century of extraordinary discoveries about the evolution of life.

Mutations: The Motor at the Heart of Evolution

To describe the fortuitous processes that lead to modifications of living organisms, we first have to specify the presence of an internal motor of evolution. This is an entity that can cause changes in the genetic information carried by DNA. In 1953, James Watson and Francis Crick discovered the structure of DNA and its code, for which they received the Nobel Prize.[17] Subsequently, many other Nobel laureates have uncovered the secrets of the genetic code, how it is read and copied, the mechanisms by which it is used to manufacture proteins that execute the physiological functions of life or serve as the architectural basis of the cell.

Thus, it is essentially thanks to the discovery of DNA, the molecule that carries genetic information, that the detailed mechanisms of life were able to be described, analyzed, and understood at the end of the twentieth century. Like Van Leeuwenhoek, Watson and Crick opened another dimension in the understanding of life. It was a new stage of abstraction in the understanding of life, because biologists did not "see" the molecules of life except by the intermediary of chemical reactions; that is, they deduced their structure and function. And we are far from having completed the analysis of this level of the details of life.

Cell biologists (whose field of observation is the cell) and molecular biologists (who study molecules) have reduced the understanding of life to extremely small structures. They have united in a new discipline: reductionist biology.[18] This analytic approach—which attempts to explain life by dividing it into component parts (a program advocated by René Descartes [1596–1650] in his *Discourse on Method*[19])—has allowed great progress in understanding how life works.

The information contained in each strand of DNA always ends up being copied and transmitted from cell to cell. The conservation of information, from generation to generation, is one of the essential properties of life. Life is fundamentally thrifty.

But copying can induce errors—these are mutations.[20] They occur in the DNA of bacteria as in the DNA of animals and plants. Advantageous, deleterious, or neutral mutations are conserved by being transmitted to descendants. DNA, the embryo of all life, can be compared to the hard drive containing the memory and the operating system of a computer governing the development, functioning, and reproduction of each cell. Then mutations, advantageous or neutral, can be analogized to additions of information, augmentation of the memory, or improvements in the operating system; they are what changes life. These mutations—copying errors—happen irrespective of the effects they might produce.

At first, biologists thought these mutations were solely random events—they were simply totally unexpected changes in genetic information and could occur anywhere and any time.[21] Of course, if we take a very long view, we should see them as contingent, as the chemical modification that constitutes a mutation would not have occurred without previous changes in the internal conditions of the cell, such as temperature and pressure, and external inputs, such as ionizing radiation. However, as these previous modifications can rarely if ever be specified, mutations appear to arise for no reason, out of the blue. There is also a contingent aspect of which mutations survive, as a particular mutation in a given environment might be favorable if some other mutation has occurred previously, but would otherwise be disadvantageous.

Cell memory adds new information or programs with no rhythm, in sudden, inexplicable fashion. These genetic mutations can be reflected by perceptible changes either of small amplitude (micromutations[22]) or of revolutionary scope (macromutations[23]), which are then exposed to the laws of nature. Some are viable, others neutral, and most are lethal to their bearers (because they lead to death, they are eliminated).

This whole process makes us think of a machine producing changes flaring up episodically in unexpected and unpredictable fashion, induced by various external circumstances. The novelties appear suddenly, modifying some morphological trait, function, or behavior of the organism. It is impossible to establish a law of probability for the appearance of these mutations. In this instance, the productivity of the machine (the appearance of evolutionary changes) appears random, though each change is contingent on external occurrences.

In the 1960s, the Japanese researcher Motoo Kimura showed that there also exists a constant modification of genetic information at the level of the molecules carrying that information.[24] In this case, either the mechanism is actually programmed to make a certain number of copying errors, or it is simply not precise enough to generate perfect copying. It operates like a scribe, a bit nonchalant in his work, who copies parchments, *always* with minor, excusable faults. Kimura therefore demonstrated that genetic information contains within itself a propensity to change through time (from generation to generation) at a constant rate. Regularly, some molecular modifications, with effects that are most often neutral and invisible, are adopted (that is, transmitted or inherited).

Most of the time, one letter more or less in a word does not change its meaning, and still less, the meaning of the sentence it is in. But repeated errors, added by chance, can form new words. At the level of genetic information, these new words sometimes produce a new characteristic for the individual that carries them.

As molecular mutations keep arising, their accumulation leads to changes in the organism. The appearance of molecular modifications and the transformations of characters they induce are therefore quite regular. Kimura thus discovered an evolutionary rhythm that was baptized the *genetic clock*. But, just as for changes caused by mutations that occur owing to external factors (mutagens, for example), they have no predetermined goal. These regular mutations, and the changes they produce, are always submitted to natural selection.

In this case, the living machine runs at a regular speed, producing novelties at a constant rate that are tested by exposure to environmental regulations. These transformations remain unpredictable and unexpected in the sense that we never know which ones will appear, and where. Only the rate at which they arise is predictable. So we can say that the productivity of this machine—the production of evolutionary changes at a constant rate—seems to be a matter of chance. These random changes can be maladaptive; they can even lead to death. Organisms have no will power, no goal; they do not plan evolutionary change. In this sense, as far as living organisms are concerned, the mutations occurring at a regular rate appear randomly, and the appearance of a favorable one is a matter of blind luck.

Today we understand that evolutionary changes produced by mutations appear either suddenly or at a regular rate (by chemical mechanisms that are either random or contingent) and that they affect morphological, physiological, or behavioral traits that are unpredictable. Mixed together in common parlance, chance and contingencies in the appearance of changes are both referred to by the vague term of *random*. Overall randomness is thus a feature of evolution. It is therefore either regularly or willy-nilly that the genetic information of every lineage changes from generation to generation by mutations.

Sexual Mixing of Genetic Information: The Second Motor of Evolution

To the contingent, fortuitous mutations are added the genetic mixing produced by sex. At the time of reproduction, of the marriage between the chromosomes contained in the sperm and egg, the information of the male mixes with that of the female. This genetic mixing can erase, change, or retain the new traits that the mutations have produced.

The understanding of this second means of modifying the information carried and expressed by each individual began with the discoveries of the Czech monk Gregor Mendel (1822–1884).[25] Though a contemporary of Darwin, he remained unknown in his time. Observing cultivated peas of different shapes and colors, he was able to establish the rules of transmission of traits from generation to generation. Every student has to learn Mendel's laws of transmission of characteristics, with diverse examples of traits (for example, blue or brown eyes) of which certain ones are dominant and others recessive.

A half-century later, researchers succeeded in explaining these laws of heredity by discovering chromosomes and their clever crossings over.[26] These occur during a process that halves the number of chromosomes that arose from the union of the chromosomes from the male parent and those from the female parent (in humans, all cells in the body have 46 chromosomes, made up of 23 that came from the father and 23 homologous ones that came from the mother). This mechanism is called meiosis, or reduction division (because of the reduction of the number of chromosomes in the cell). I described this in chapter 4 as a divorce, a reconstitution of simple cells that have only one set of chromosomes and that are destined to become the sex cells, or gametes (in animals, these are the eggs or sperm). This mechanism is in fact much more complex than a simple division into two similar sets of chromosomes. During meiosis, we observe:

- An exchange of segments between two homologous chromosomes. This is a true mixing of information within each chromosome (intrachromosomal mixing). For this to happen, during the course of meiosis, homologous chromosomes (in each pair, one from the male parent, the other from the female parent) pair off, line up, cross, and exchange small portions (this is *crossing over*).
- A random association of chromosomes coming from the mother and those coming from the father. This is a matter of mixing the chromosomes in a set (interchromosomal mixing). During meiosis, after crossing over between homologous chromosomes (one from the father, the other from the mother) allows the exchange of segments, the chromosomes separate from one another and partition themselves randomly into two homologous sets (each set constitutes a new information package carried by eggs and sperm). The different eggs and sperm therefore have different percentages of chromosomes coming from the mother and from the father of the individual that produced them.

The rules of probability apply to these mixings, leading to the transmission of many traits.[27] The recombination or reassortment of chromosomes and of their fragments can cause death, can be neutral, or might constitute an advantage for the individual that carries the new set.

For this to happen, a male and a female must meet. Chance and contingencies govern the probability of these meetings (except in humans and the species we cultivate, because we understand the genetic process and can manipulate it to our advantage with planned matings; the transformations we accomplish are no longer random but are indeed expected).

Mutations and mixing of genetic information complement one another and are responsible for all observed evolutionary changes. These two distinct mechanisms produce modifications that are then tested by natural selection.

To describe the characteristics of these modes of transmitting information, with their errors, mixings, and exchanges, scientists use printing terms: replication, transcription, recombination, transposition, translocation, translation, fission, fusion, alteration, substitution, deletion, repetition, reshuffling, inversion. These words apply to parts of the "book" (in this case, the nuclei) that constitutes the totality of the information of life: chapters (or, chromosomes), pages (parts of chromosomes), paragraphs (genes), lines (sequences of nucleotides), words (triplets of nucleotides), and letters (nucleotides). And we must bear in mind other fragments, other phrases, so to speak: genes, introns, exons, codons, transposons, sequences, nucleotides. Biologists even distinguish highly informative codes (chapter titles) and codes lacking information (spaces, blanks).

Mutations and Sexual Mixing: Contingent or Predestined?

Genetic information governs the functions and forms of life; it conserves the past while introducing and capitalizing on changes from one edition to the next. The detailed, impressive investigation of life has not led to clear new laws allowing us to model evolution. The overall conclusion is humble: the accumulation of mutations and the sequence of chromosomal mixing constitute a fortuitous chain of events that occur totally independently of one another; they are contingencies.

These mechanisms for changing genetic information go on all the time. The neutral or advantageous changes accumulate automatically. Physiological and morphological change cannot be avoided. In changing environmental conditions, species that do not adapt end up being eliminated.

This whole system has inspired many thinkers who ask about control of such a dynamic system. If change is automatic, evolution is neither disordered nor contingent. Who wrote the program for the big game of life?

Interpretations of the essence of life unrelated to scientific considerations have prevailed from the time of the first prophets of the Jews and the Greek philosophers through the twenty-first century. Theologians and philosophers have long argued that only a divine force—a Great Architect—could have created the final result: nature, a harmonious ensemble that we find beautiful and preside over, thanks to our self-consciousness and our spirituality. In general, these interpretations are all in accord in considering nature as frozen in time in a state predestined for man. Even the Eastern religions consider nature as a harmonious ensemble of divine creations or representations of the divinities themselves. For Confucianism, animals, plants, and even inanimate objects derive from the same vital energy that created Man, and so they all deserve the same respect.

Religions and philosophies have also had to evolve. As science has progressed, the notion of change in species was slowly considered, then admitted by the Roman Catholic Church. In the 1870s, Pope Pius IX (1792–1879) did not hesitate to excommunicate priests daring to refer to Darwin's work. Seventy years later, Pope Pius XII (1876–1958), in his encyclical *Humani generis,* proclaimed that the theory of evolution was a hypothesis that ought to be considered with caution and moderation, always subject to the judgment of the Church, while emphasizing that the creation of souls remains a divine process. He pointed to scientific uncertainties about evolutionary theory; he doubted.

In 1955, the Jesuit priest, philosopher, and scientist Pierre Teilhard de Chardin[28] succeeded in taking into account the whole body of knowledge on evolution in a spiritual exploration similar to the one adopted a half century earlier by the Nobel laureate and philosopher Henri Bergson.[29] For Teilhard de Chardin, contingencies or probabilities leading to transformations during the history of life were systematically biased or rigged by directional forces, a "vital spirit" or sort of anti-chance that predestined change to occur in certain directions. The evolution of life, in this view, was fated to follow a constant direction (he called this process *orthogenesis*). Similarly, to distinguish man from the animal lineage he sprang from, Teilhard de Chardin invoked a divine genesis of the mind (*noogenesis*).

Nearly 40 years later, in 1996, in the face of an avalanche of ever more precise scientific data, Pope John Paul II (1920–2005) sent an encyclical letter to his bishops. In this paean to research into truth and the relationship between faith and reason—"*Fides et ratio*" (faith and reason)—there is a critique of the "Biblicists" (those, like creationists, who refer only to the Bible, which is for them the only true reference). He underscored the legitimate interest in the progress of scientific knowledge and research, all the while reserving judgment on evolution. On this subject, he recalled the statement of Pius XII, though he did not affirm it; he seemed still to doubt. Then, in October of that same year, the pope made a surprising declaration in French before the Pontifical Academy of Sciences, entitled "La vérité ne peut pas contredire la vérité" ("Truth cannot contradict truth"), in which he forcefully affirmed that the theory of evolution was proven and that it was compatible with Catholic religious doctrine.[30]

Between the appearance of Darwin's *On the Origin of Species* in 1859 and this papal message, 137 years had passed.[31]

This whole account summarizes the scientific and mystical history of knowledge or interpretations of the ancient pathways of life.

The Goal of Life

Though present-day science enlightens us on the causes and effects of evolutionary change, there remains a classic question that all naturalists, philosophers, and theologians have posed from time immemorial: this concerns the future, the goal of life, the why.

Is life an end in itself?

It is remarkable that all religions have responses to that question that have essential recourse to spirituality.

The Judaic, Christian, Muslim, Shinto, Taoist, and Confucian religions define both a mystic origin and a predestined finality: rebirth in paradise or in another life on Earth. For Teilhard de Chardin, two forces—ascendant evolution and a spiritual energy—join to produce a destiny in which humanity will be united in a mystical eternity with Christ.[32] It is important to note with respect to this theme the unique take of the Church of Jesus Christ of the Latter-Day Saints (the Mormons), which sends tens of thousands of missionaries to traverse the planet to certify deaths in order to baptize dead individuals so they can reach heaven.[33] Islam, for its part, magnifies paradise. Its richness, splendors, and delights are described in the verses of the Koran. Poor and rich are greeted there by angels and never know poverty; this home is eternal, people do not die there, clothes are not worn there, and people remain young forever.[34]

In Asian religions, life is recyclable. Reincarnation, also called the wheel of existence or the law of karma, is the driving element of Buddhist thought. The vehicle of Buddhism corresponds to the one the Buddhist borrows to cross the river of reincarnation and reach Nirvana. The supreme goal in this religion is to escape from the wheel of reincarnation and attain Nirvana, or, illumination. One achieves this state by withdrawing from the material world and consecrating oneself to eliminating desires and actions.

Scientists have addressed spiritual concepts of the finality of life. The French biologist Jacques Monod (1910–1976), Nobel laureate and codiscoverer of the mechanism by which genes turn on and off to produce proteins, was the first to consider on a philosophical or metaphysical level the discoveries on the transmission and transcription of genetic information.[35] The scientific insights brought to the subject by Monod and other reductionist biologists led to a definition of a materialistic objective for life. Life has only one goal, in this view: *to multiply itself ad infinitum.* In the structures and functions of life, all organisms strive towards this mechanistic reproductive goal free of any humanistic or religious significance. In this sense, François Jacob (with Monod, a Nobel laureate in medicine) stated the unique dream of every cell, which is simply to reproduce.[36] According to these biologists, living entities are distinguished from nonliving ones by the fact that they have an activity: to perpetuate themselves indefinitely. This project, oriented and coherent, is constructive, even though the means of its application cannot be foreseen in specific cases.

This idea was taken up again six years later in 1976 by the English biologist Richard Dawkins,[37] who described a chemical automaton. According to this idea, everything began with a primer, a molecule similar to DNA, which formed by chance. This primer was able to perpetuate itself forever, thus constituting a sort of chemical memory. This is the only end! For Dawkins, *all life is only the temporary bearer of genetic information, and its reproductive tool.* Several lines of evidence suggest that this primer was RNA, a single-stranded molecule very similar to DNA, which is double-stranded.[38]

For Dawkins, the genes, bearers of information, are chemical memories that never stop propagating, and whose selfish destiny is beyond human reason and is in no way a mystic, programmed endpoint or goal. This concept, called the "selfish gene," is based on the endless duplication of DNA, the molecule carrying genetic information. An illustration of this idea is found in the fact that certain DNA sequences, apparently with no function in the organisms carrying them, can nevertheless transmit themselves endlessly to future generations. This selfish DNA, with no *raison d'être,* is a sort of ultimate parasite that survives from generation to generation inside the bodies of many ephemeral living individuals.[39]

This approach goes so far as to propose explanations for certain behaviors that do not seem selfish. Therefore, in this view, an animal's acts of altruism, just like many sexual behaviors, always have a hidden goal: to assure, above all else, the perpetuation of the genes, even to the point that the individual can sacrifice itself to this end. This universal law of selfish transmission of genetic information is blind and lacks any sentiments; it also explains many behaviors seen in nature that seem to us to be gratuitously cruel.[40]

The philosophical theories of Monod, Jacob, and Dawkins converge in identifying a goal for genetic information: that of having an endless destiny whatever happens, whatever the habitat the living organism finds itself in, whatever the environment it inhabits. This materialistic objective is the only destination of the pathways of life.

These considerations, deprived of any spiritual aspects, imply that the present composition of life has no meaning. Its apparent harmony is only a fortuitous effect. All living organisms are only in essence packets of information competing with each other with a single goal: to reproduce. In this sense, the extraordinary structured organization of the diversity of life that surrounds us today is only a momentary image resulting from the contingencies, lucky and random events in the transmission of genetic information used in pursuit of the goal of eternal reproduction. The harmony of nature is relegated to a succession of temporary, unstable states that constitute fragile equilibria among all the organisms competing to transmit their genes.

Natural Selection of Mutations and Sexual Recombinants: The Pilot of Evolution

What remains of the mechanism of natural selection proposed by Darwin, who knew nothing of Mendel's laws, of chromosomes, of genes, of DNA? For him, evolution can be summed up in three simple observations, completely independent of the abstract considerations of reductionist biologists:

1. Species change, and some fraction of these changes is transmitted to offspring (we should not forget that Lamarck laid down the bases of "transformism" of species 50 years before Darwin).
2. Organisms produce more offspring than can survive.[41] There is thus a triage, a continuing elimination of certain offspring and survival of those that are better adapted.[42]

3. Offspring that change in a way favored by the environment will survive and reproduce better. Such changes will then spread (increase in frequency) in successive generations. This is change by natural selection.

The role of the rigorous, implacable environmental exigencies in determining the main directions of evolution was Darwin's essential insight. Has the role of these external constraints been marginalized by later discoveries about mechanisms of transmission of genetic information within cells (mutations and chromosomal mixing)?

Not at all!

It is a question of another level. Darwin explained the reasons for necessary changes of species through time. He accurately identified the constraints governing these changes: environmental conditions gradually mold structures, forms, and functions of the lives that surround us. The natural environment is the master sculptor of life. This is how Darwin established his theory; the motor of change is natural selection, even if he did not know the fuel of evolution (the fortuitous, contingent modifications of genetic information by mutation and sexual union).

Therefore, over the various sorts of modification of genetic information that supply the malleable material is superposed a selector, a master architect of forms and functions of life. It is a matter of the physical, chemical, and biological conditions in which each species develops. Life evolves in shackles imposed by the external forces that determine, on the one hand, the most logical forms in a given habitat[43] and, on the other hand, the functions best adapted to the environment of an organism (those that most favor its growth and reproduction). Physical and chemical constraints plus natural selection pilot the machines that produce the changes (the accumulated, ever-changing genetic information). They raise barriers beyond which the wandering of changes induced by mutations and sexual reproduction cannot go. They orient the paths of life, but in a blind fashion, with no plan or goal. Richard Dawkins constructed an entire book around this theme, comparing natural selection to a blind watchmaker.[44]

Whether the project of all life is to transmit genetic information or to attain some mystical destiny, it is always true that the information transmitted from generation to generation is molded and selected by external constraints. The project of infinite, selfish duplication of genetic information therefore seems aimed at an unknown goal.

Filter, sieve, mold, collar, corset—biologists use all these terms to characterize the role of external constraints in natural selection.[45]

Natural Selection: A New Contingency

Considering the evolutionary importance of the constraints of the physical environment, theologians and some philosophers can still perceive a determinism, a programmed fate, a finality—fruits of divine selection. For them, mystical forces orient the shaping of life and of natural selection in order to produce the harmonious nature that we observe, including our own superiority within it. In this conception, it is this guidance that replaces the chemical machinery. A divine force guides the *materialistic* project (mechanical, automatic, blind, and selfish) of eternal duplication. The external forces shaping life and natural selection are *deterministic;* they allow the necessary adaptations of organisms to local conditions. There is a finality in the destiny of life.

But examination of the ways exterior forces act on the elaboration of life and on natural selection shows that they are not at all deterministic. There are mechanical constraints (physical and chemical) that lead to self-organization of logical forms (these are the same constraints that govern the formation of crystals), and natural selection is materialistic in the sense that it acts by a sequence of fortuitous events linked to one another. Contingencies! In fact, the evolution of life, powered by natural selection, is essentially contingent, resembling a lottery game in which winnings are never withdrawn. There are no fixed probabilities for whether new, promising changes or ideal forms, products of genetic mixing and transcription errors, will be validated or eliminated by nature. They must arise at the right time and place. The variants of finch beaks that Darwin saw could as well have arisen genetically in the continents the finches originally came from. But there they never would have been favored by selection, because they would have conferred no advantage there. By contrast, in the very different context of the Galapagos Islands, where the ancestral finches arrived purely by chance, the successive transformations leading to the appearance of differently shaped beaks adapted to local foods were selected for. In other words, for favorable heritable changes to appear and be adopted, luck and the contingencies of the internal transformation of genetic information are key, and the circumstances of the external environmental situation are equally crucial. Natural selection is therefore both fortuitous (everything rests on the environment that confronts the genetic modifications) and precise and deterministic if we are referring to the way selection works in a particular environment; the best-adapted forms always win out. The great watchmaker of natural selection always chooses the best mechanism for a given context but does not know (or cannot see) that, all the while, he is constructing a watch.[46]

The contingent and chance games of genetic transformation, internal in all plant and animal cells, are also submitted to another test, that of contingencies of the external environment in which the organism evolves. These contingencies, the changes in an environment that affect the species living in it, drive evolution by spurring sequences of transformations.

On the scale of an individual human life, these contingencies are everywhere. Our birthplace, the makeup of our family, income and educational level, the country and time in which we live—all these environmental circumstances guide our destinies. Let us take a classic example, that of a person known for remarkable artistic ability. We say that she is gifted in a particular art. Even though the heritability of such a gift has never been established (it is not transmitted according to Mendel's laws), on the individual level, the existence of an innate trait for drawing or music is often assumed. Certainly, by dint of work and technique, one can acquire artistic skill (this is an acquired trait, never transmitted), but among apprentices, certain ones will clearly be more gifted or sensitive than others in expressing themselves (this would be an advantage inherited by chance). If the artistic gift seems well shared among families of musicians or painters, it is simply because of an upbringing in which the trait is fostered, so it is more likely to be manifested. Put another way, there are environmental circumstances that favor the expression of the possible gift.

We can imagine that Vermeer, whether by contingencies or luck of genetic mutations and recombination, must have been born with a gift for painting. He lived in a beautiful city in which art was admired, he was raised in a family of art dealers, and he profited by his marriage from the patronage of a wealthy family inclined towards all forms of artistic expression. This context was favorable for the expression of his innate gift. It was a succession of events with a fortuitous outcome! He could have been born on a farm or married a peasant from the rural area around Delft; his gift would doubtless have been left unexpressed in the course of a life consecrated to caring for Dutch cows.

It is the simple principle of environmental contingencies, which also applies on the level of the evolution of species by various means. In order to explain the diversity of plant and animal life that surrounds us, in addition to the inevitable contingencies and random lucky events of transcription and transmission of genetic information, we must therefore add the contingencies of environmental circumstances encountered by an organism that carries new information. It is a long chain of fortuitous events.

In the first part of *On the Origin of Species,* Darwin suggested the existence of environmental contingencies by referring to the artificial selection practiced by breeders. His example had to do with the breeding of pigeons. To select high-performance pigeons, the breeder experimentally eliminates the contingent environmental events by raising the desired pigeon in optimal conditions. Moreover, he, rather than nature, is the one who chooses the pigeons that manifest a desired set of traits. He crosses the pigeons he has selected and that have grown up in an artificial, ideal environment. The sought changes are quickly manifested from generation to generation. Breeders are thus effecting a directed selection; they place an exceptional individual in a favorable environment and cross it with an individual with the same traits. Subsequently, the most promising descendants are again crossed with one another, and this process goes on for generations. Obviously in this context it is not nature that sculpts the lineage, but the breeder. But the principle is the same. The changing clay constitutes variation of internal origin—genetic—that each individual contains within itself. Without the breeder, nature can still work to cause the lineage to evolve, but the appearance of a pigeon that is superior (according to the standards of the breeders) is then the fruit of a series of fortuitous contingencies.

By suggesting that the contingent conditions of nature actually cause the transformation of species, Darwin struck a strong blow against belief in a predestined nature designed by God for man. Darwin understood the heretic nature of his observations. He lived in an age when his conclusions were very politically incorrect, and therefore he avoided emphasizing the importance of environmental contingencies.

Promoters of Communism also understood that Darwin's work included antireligious principles that were in harmony with their revolutionary, materialistic theses. This was why Karl Marx (1818–1883) sent Darwin a dedicated copy of his book *Das Kapital.* Because he could read German only with difficulty and was not very interested in politics, Darwin read only the first part.[47]

In the history of sciences, the discoveries I have just described actually occurred in reverse order. Scientists first granted that species can change. Next, they recognized that natural selection plays a preponderant role in these changes. Finally, they understood how these changes were initiated by mixing of chromosomes and changes in genetic information. All these discoveries have subsequently been integrated into a synthesis, rich in nuances and details of particular cases, that is endlessly improved by research.

From Theory to Theorem

In the time since *On the Origin of Species* was published, the evolution of animal and plant lineages has become a rather trite theme. The portrait of Darwin as an old, bearded, forbidding figure symbolizes for most people the man who opened our eyes to natural history. Though the concept of the transformation of species (that is, the evolution of life) traces to Lamarck, the description of the main mechanism of evolution will always have Darwin's writings as a starting point.

However, if we plunge into the abundant literature on evolution, we cannot fail to find the works of a prolific American author who contributed greatly to popularizing these themes during the last few decades and to whom I want to render homage (I often cite him in this work, and he died shortly after I wrote these lines). He was a professor at Harvard University, where he occupied a university chair. He wrote on geology, paleontology, evolutionary biology, and the history of science and was a key contributor to many seminal papers in evolutionary theory. This is Stephen J. Gould. For 25 years, he regularly wrote articles for the popular science magazine *Natural History* and published collections of them as books.[48] His original presentations on natural history are packed with diversions, autobiographical notes, and metaphors. His style of presenting an aspect of natural history by beginning with a seemingly insignificant, irrelevant anecdote, yet ending with a fundamental treatment of some evolutionary matter, is elegant and pedagogically effective. For a biologist who wants to begin writing a popular account of present-day evolutionary thinking, it is frustrating: this major evolutionary theorist said so much, dealt with so many topics. Such a degree of preeminence cannot help but arouse jealousy and objections.

Jealousy on the part of molecular biologists and geneticists who could not fail to observe that he popularized natural history so well without sufficiently integrating their reductionist sciences to suit their tastes.

And objections on the part of religious fundamentalists, partisans of biblical creationism, who recognized in him a brilliant adversary who knew exactly how to choose and to present, among the jumble of information on natural history, so many cogent proofs of evolution. A master of academic jousting, Gould played a determining role against these adepts of a literal interpretation of the text of Genesis. These creationists wanted—and still want—to impose a presentation of the diversity of life solely by reference to the mystical events of the seven days of creation in the biblical narrative. Gould was a valiant warrior in public debates and lawsuits on this peculiar debate, more typical of an earlier era, in a nation where science is so well developed. In these battles, which he often wrote about, he certainly became more assured and determined to communicate his views. I greatly admire him for having had the audacity to leave the ivory tower of a comfortable university where the only respectable communication is high level scientific publication that only other scientists understand.[49]

In this book, I often refer to the writings of this disciple of Darwin to reconstruct, in my own way, my synthesis of the evolution of animals and plants.

Of course, it is true that the revelations of Van Leeuwenhoek, Lamarck, Darwin, Mendel, Watson, Crick, Kimura, and so many others, as well as the metaphysical interpretations of Monod, Dawkins, and Gould, have contributed to the demystification of life and its removal from the sacred sphere. Nevertheless, they should not be ignored by believers. On the contrary, their work can be used to orient our ethics, our behavior in the face of the lives that surround us. Whether present-day organisms have been created or "designed" by God or whether they are cleansed of all mystical origin (and are the fruit of random fortuitous events and contingencies), we can all admire, respect, and protect them. And if science allows us to protect life better, why not use it?

In this sense, science and religious belief can coexist, without conflict and without ignoring each other.

In the United States, it is remarkable to observe the permanent linking of contradictory, Manichean works based on ancient stories: some defending the religious views inspired by Genesis (including the concept of intelligent design), the others referring essentially to Darwin (the revered symbol of evolutionary theories).[50] Scientific fundamentalism is exemplified by a book by Richard Dawkins; stressing the evidence on the importance of chance in evolution, he preaches a hard-line atheism.[51] He even provides an appendix in his book that lists atheist associations that can help believers to convert. Stephen J. Gould participated in this debate on the opposite side of creationists during his whole life as a researcher. Bad faith and dogmatism have hardened the confrontation between two diametrically opposite conceptions of the origin of life. Gould offered a proposal to lessen the tension; he called it NOMA—the non-overlapping magisteria.[52] According to Gould, religious and scientific magisteria should remain independent, working in their own domains without pronouncing on subjects that they do not both deal with. In other words, science (which is objective) should not criticize spiritual interpretations (which are subjective) of knowledge, and conversely, religion should not impose dogmas in place of uncontested scientific data. The result is a calming status quo. Each in its own sphere, with different roles, but the search for truth is common to both. This is Galileo acquitted without having to recant, free to defend his scientific ideas but not to comment on the spiritual implications of his discovery. Heliocentrism is a subject common to both magisteria, so it can be debated between them. Objective truths have shown that the Earth revolves around the sun. This fact can be adopted by the two magisteria, while its spiritual interpretation is left to the magisterium of religion without scientific objection.

Personally, I find that this peace, based on respect of beliefs and on progress in and recognition of knowledge, serves humanity. The spiritual is respectable and scientific knowledge can be integrated into it; the essential point is harmony between the two spheres. No more spiritual dogma concerning verifiable facts (there are, for example, the dogmas of biblical creationists whom Gould battled), no more scientific doctrines on spiritual thoughts.

The Catholics Teilhard de Chardin and Pope John Paul II, the associations of theologians of the liberal Protestant churches, and certain Koranic schools have integrated, each in its fashion, the extraordinary advances in scientific knowledge of the last century. To convince yourselves of this fact, read this quotation, then look up the author in the endnotes:

> Today, . . . new knowledge has led to the recognition of more than one hypothesis in the theory of evolution. It is indeed remarkable that this theory has been progressively accepted by researchers, following a series of discoveries in various fields of knowledge. The convergence, neither sought nor fabricated, of the results of work that was conducted independently is in itself a significant argument in favour of this theory.[53]

To be completely frank, for someone who does not live in a nation in which religious doctrine serves as law or in the United States, where the scars of virulent opposition between creationists or those who believe in intelligent design and scientists are still raw, the problem of teaching the genesis of life has disappeared. In Europe, Catholic, Anglican, and other

Protestant churches have not made teaching evolution a *casus belli*. I even think that, because the debate has ended, the teaching of evolutionary theory has been a bit neglected. It consists of the history of science, natural history—not very fashionable in the context of reductionist themes in biology! It is unbelievable, but at the University of Nice–Sophia Antipolis, when I mentioned the mechanisms of evolution in my ecology course for students in their first year of biology (I did not teach evolution courses before 2004), I was surprised to find that the great majority of them had difficulty stringing together two sentences on Lamarck's and Darwin's ideas on evolutionary mechanisms.

Religions, philosophies, and ideological currents can be respectable and salutary. Spiritual, moral, and political values, supported by rites, symbols, traditions, languages, and doctrines, usually legacies of a distant past, unite men and women, communities, and citizenries in certain lines of conduct or shared aspirations. Biological progress is not really opposed to these values that are so useful for humanity; these values can even enrich scientific advances related to evolution. Understanding of the history of life grows from instruction and knowledge that endlessly change, improve, and progress. Such knowledge is increasingly necessary for better evaluation of our misdeeds with respect to nature and to improve the future for our descendants. Thus, in the disastrous dynamic of the growing domination of man over nature, it is dangerous for humanity to impede the integration of its scientific discoveries into common knowledge and policy. The obscurantism of certain ossified doctrines, now outmoded, serves only to vilify individuals or to widen the abyss between science and the most committed fundamentalists.

By the same token, the peace of the magisteria should not leave the field free to opportunistic gurus who elaborate theses that are a mélange of astrology and fatalism with phony glosses of cosmology and DNA. In this category I would place the pseudoscientific prophets of so-called intelligent design, who have achieved a degree of political success in associating scientific facts (the stages of evolution) with a religious belief—that of a superior force that alone can guide evolution toward the extreme complexity of modern day life, crowned by the creation of man. It is desirable to integrate scientific knowledge into spirituality, but to use scientific notions as a veneer to embellish mystical doctrines is lamentable. If sects want to impose a spiritual obscurantism based on pseudoscience with the goal of rallying the faithful, they are breaking the peace treaty of NOMA. It is not for biologists to start a war of ideas, but they have important information. They also have a responsibility to communicate it. They should freely share their knowledge and discoveries even on sensitive subjects that will annoy some spiritual communities who mislead their followers by relying on erroneous science to further their beliefs. To unmask the scientific ineptitude of charlatans and to underline the distinction between science and spirituality in a story about evolution are also parts of the legitimate role of those who advance knowledge.

One day, in my laboratory, I had to identify an alga under the microscope. I had just drafted this chapter and I was still dreaming about all the philosophical and theological fallout from progress in biology. Around the alga swam minuscule cells of different shapes and colors. Admiring these little, moving creatures, I thought that I would have loved to discover them, describe them, exhibit them, and interpret them. I put myself in Van Leeuwenhoek's place. There's one, quick! I have to draw it, compare it to the others, understand its life cycle. That must have been incredibly exciting! He must have peered into every liquid in nature, looking for life. He must have gone from surprise to surprise.

Looking at the microscopic moving cells, I saw just beneath the surface the two Vermeer paintings in which the young Van Leeuwenhoek brushes a celestial globe or measures a map distance with dividers. I imagined him distinguishing these strange creatures with the aid of his primitive microscope. He surely marveled at their diversity.

Atomic physicists have an emblematic master: Albert Einstein (1879–1955).[54] In one photograph of him he is sticking out his tongue. This is a symbolic image of supreme knowledge associated with human simplicity. For me, to remember the illustrious biologists of the past, I prefer the serene portraits of Van Leeuwenhoek painted by Vermeer to the widely printed photographs of Darwin's worried, austere visage or those of the ordinary silhouettes of Crick and Watson.

Vermeer's *The Astronomer* is exhibited at the Louvre in Paris. If you have occasion to see it, observe the delicate gesture of his right hand brushing the celestial globe, which represents the infinitely large and is a symbol of scientific knowledge. The position of his left hand is, on the contrary, resting on a table of banal, everyday things. You should also see behind Van Leeuwenhoek a representation of Moses saved from the waters. It is important to recall that the infant Moses was set adrift in the Nile in a wicker basket to escape the terrifying decision of the Pharaoh to exterminate all Jewish male children. He was saved from drowning by Pharaoh's daughter, who adopted him. Ever since this story has been told, it has been interpreted as an allegorical image of reconciliation between two enemies.

Therefore, I see in this painting the reconciliation of the spiritual and scientific magisteria. Vermeer thereby personalized forever, with grace and refinement, mankind perceiving simultaneously the infinitely large heavens and the microscopic ferment of animal and plant life.

To the abstract discoveries of the cellular and molecular dimensions of life must be interwoven discoveries on its evolution. Genetic mutations and mixing plus natural selection constitute a series of three independent events, lucky or contingent. This mechanism has sculpted life at each instant, taking account of the physical and chemical forces acting at the time, the habitat, competitors, and experiences of the past. Together they explain how species change through time, how they evolve. Lamarck's aphorism, "species change through time," in fact describes evolution, and it is no longer simply a theory, nor a hypothesis surrounded by uncertainty. It is a proven fact, a biological law. "Life evolves" is a theorem—not a theory!

In the evolution of plants and animals, everything began with prototypes consisting at the outset of single cells.

CHAPTER 6

The Densimeter

The First Flourish of Animal and Plant Life: A Historical Inquest

Was Vermeer's model for *The Astronomer* and *The Geographer* really Van Leeuwenhoek? The more time passes, the rarer the discovery of new elements from the archives of their era; the sources of evidence are drying up. We must, therefore, tentatively accept the best hypothesis accounting for the accumulated information. In this particular case, several indications all lead to one conclusion: Van Leeuwenhoek was indeed Vermeer's model.

The Four Sources of Evidence for the Reconstruction of the History of Life

All historians confront this sort of uncertainty because of the erosion of evidence by time. Four collections of distinct sorts of data guide the reconstruction of the history of life.

The first collection consists of the forms, anatomies, cells, functions, and behaviors of the millions of species that constitute biodiversity today. All these features characterize life. Beginning in the late eighteenth century, they have allowed us to name present-day species. These are classified according to differences or similarities that most often indicate real genealogical relationships.

The second mass of data is composed of the genetic information of life. This data source has been exploited more recently (beginning less than half a century ago). Its basis is a fact: each organism carries DNA, the material of genetic information. It can be seen as a coded text with an alphabet of four letters (each letter is the first letter of the name of the four fundamental molecules making up DNA). It has been established that the genes, similar to paragraphs of a coded text, represent the instructions necessary to build organic molecules: the proteins. The nature and action of proteins determines all the characteristics of life. DNA is also an accumulation of fossil information. In effect, each past change in the genetic code has been accumulated and saved with little alteration. This is why most of our genes are identical to those carried by all mammals, and some of them are identical to those of simpler organisms. Therefore, the analysis of DNA, with a view to reconstituting the history of life, is like an explication of a text along with a search for plagiarism. We compare the DNA of two modern species, seeking on the one hand the parts that are identical or very similar (the concordances) and on the other hand the meaning of genes (to which traits do they correspond?). This is how relationships or divergences are deduced. Likewise, successive transformations of species have a rhythm set by the regular accumulation of mutations (see chapter 5). This rhythm allows an estimate of dates of divergences (this molecular technique allowing the dating of past transformation is called the *biological* or *molecular clock*).

Using these first two data sources, we can reconstitute genealogical trees of life and group present-day species according to their phylogenetic relationships.[1]

The third sort of data is provided by fossils, which paleontologists dig up and study. The last few decades have seen great strides in techniques of dating fossils and in analyzing their constituent materials. They can therefore be considered as reliable temporal milestones to aid in reconstituting the history of life. This sort of knowledge of past life is inaccessible to the two preceding means of investigation. For example, it is not in studying the current diversity of life, or the DNA of modern organisms, that we can deduce that dinosaurs existed on Earth.

The fourth data source is related to all the previous types of information. It concerns both the environmental conditions prevailing in the past (determined by geologists) and the behavior of modern species in relation to these conditions. This is an ecology of the past—*paleoecology*. Reconstructions of faunal and floral equilibria that occurred in the distant past can be deduced by considering elementary ecological rules.

The best evolutionary scenarios take all four of these types of data into account. However, as with those who are forced to describe the life of Vermeer as best they can, historians of nature also observe that the further back in time they go the more difficult it is to discern the truth. However, if the life that surrounds us has already surrendered many of its secrets, the archives of genetic information are still being deciphered, and fossils are far from having all been discovered. The real hope that we may know more about the history of life tomorrow should not lead us to excessive prudence today for fear of being contradicted. Imagination integrating current knowledge can orient the discoveries of tomorrow.

The First Two Billion Years of the Evolution of Plants and Animals

The great majority of works dealing with the evolution of life concern species that have succeeded one another on Earth over the last 570 million years. The pathways of life during this period are increasingly well reconstituted thanks to an abundance of petrified remains. By contrast, the first two billion years of the evolution of fauna and flora (that is, between 2.7 billion and 0.57 billion years ago) remain mysterious.[2]

The chronology of the development of life in the very distant past must be studied in minute detail. We have to look for the rare reliable data that are available, then link them together in a coherent scenario. We must confront knowledge from different disciplines and keep coming back to our theme. It is a matter of the same sort of meticulous work that art historians accomplished as they patiently reconstructed the life and work of the enigmatic Vermeer. They identified the distinct stages of his life with the aid of disparate sorts of evidence, like that found in the archives of Saint Luke's Guild of Delft, a fraternity of artists and merchants where Vermeer was a master artist, or in sales catalogs or marriage licenses or birth or death certificates. They were able ultimately to give a name to the man who modeled for those two paintings: Van Leeuwenhoek. All these elements also allowed writer Tracy Chevalier to create a realistic history of the painter's life in the novel *Girl with a Pearl Earring*.[3]

An artist's notoriety is linked to the value attributed to his works, which leads historians to study his life. For the history of life, it was above all the emergence of evolutionary theories that impelled researchers to attempt to reconstitute the past of life on Earth. In this effort, several researchers of the late nineteenth and early twentieth centuries were perplexed by the problem posed by the absence of very ancient evidence of life on Earth and the sudden appearance—in rocks dated from 570 to 520 million years ago—of many kinds of animals and algae. Thus, during the half century following the appearance of *On the Origin of Species*, only three types of fossil organisms dated to more than 570 million years ago were described, and these were immediately challenged.[4] This was the reason Darwin sometimes doubted his own theory of evolution—there was an absence of a real beginning, a lack of existence of simple forms before the sudden appearance of complex ones. Some scientists defended a compromise: yes, species changed and evolved from one into another, but their genesis was so sudden that only God could have created, in one feel swoop, the original biodiversity. Everything that preceded the Cambrian (between 550 million and 480 million years ago) was therefore named the Azoic era: the period without life.

It was not until the second half of the twentieth century that many petrified vestiges of older organisms were unearthed.[5] And now, at the beginning of the twenty-first century, researchers are discovering the imprints of several hundred minuscule forms of nonbacterial organisms that lived between 2.1 and 0.57 billion years ago. They attest to an early genesis of fauna and flora.

This rarity of reliable data, extending over many millennia, causes us to be cautious. Most fossilized organisms of this epoch are formed of just a single cell. Each one cries out to the researcher: is this a bacterial cell? A cell of a prototypical animal or alga? Or a resting stage of an alga or an egg of an animal (in each case mature individuals would be multicellular)? Other questions are evident in interpreting the very rare groups of identical cells, attached to each other and all dating from the same time. Are these filaments of bacterial cells? Of algae? Of animals? Of lineages that have disappeared forever or of distant ancestors of forms alive today?

There are not enough intermediates to establish relationships.

The comparison of genetic information contained in the various species today does not allow reconstruction of the diversity of the first branches of the genealogical tree of life. Reliable genealogical reconstructions from molecular information depend on the number of species compared and how representative they are.[6] Now, present-day plants and animals do not represent all the genealogical lineages that existed hundreds of millions of years ago. Furthermore, organic matter is preserved for just several millennia; most ancient DNA is highly degraded after 100,000 years.[7] Therefore, the genetic information carried by the majority of the hundreds of millions of extinct species that have marked the history of life will remain forever inaccessible,[8] meaning that genealogical reconstructions from genetic information of living species are biased. This information bears only on the distant phylogenetic relationships of *present-day* species and on the approximate dates of past divergence of *their* distant ancestors.

These considerations make us understand that after each hypothesis for the reconstitution of the first five-sixths of the existence of animal and plant life, we have to wait for a falsifying fact from some new discovery. However, in the context of great uncertainty, the magnitude of recent discoveries on the remains of life of this mysterious epoch has led to synthetic focusing of hypotheses in which imagination and intuition play a large role (scientists refer to such cases as speculation). In such situations, it is no sin to err. In this historical development of thought the experimental sciences give way to simple deductions. The value of scenarios rests only on their logic in the face of reliable but disparate temporal data. This is the exercise I am undertaking here.

At the Dawn of Life, Animals and Plants Were Microscopic

To begin, we must draw up the balance sheet of the situation before the dawn of unicellular animals and plants. In chapter 2, I laid out the hypotheses on the origin of life and referred to data on the most ancient traces of life, represented for nearly a billion years only by bacteria (there are contested discoveries of bacterial molecules dating from 3.8 billion years ago and fossil bacteria dating from 3.5 billion years ago; increasingly frequent, uncontested discoveries of bacterial remains date from between 3.5 and 3 billion years ago).

Among these fossils, paleontologists have identified "plant-like" bacteria that owe their lives to light (represented by microscopic independent cells or filaments of identical cells). Well before 2.7 billion years ago, they became so numerous and so widespread in all the seas and oceans that they ended up oxygenating the upper layer of the waters they inhabited.

Amidst these bacteria evolving in oxygenated surface water, the genesis of prototypes of plant and animal cells occurred.[9] In chapter 4, I described how these prototypes acquired mitochondria, ex-bacteria allowing respiration in an oxygenated environment. I also explained how animal and plant life invisible to the naked eye (unicellular) arose from the fusion of bacteria. This genesis, by addition or fusion of bacteria, could have begun more than 2.7 billion years ago.

This hypothesis rests on one observation. Biochemists have discovered that in natural organic products of animal or plant origin that are rich in carbon and hydrogen, found in Australia and dated to this time, there are molecules that only organisms other than bacteria can produce.[10] The prototypical animals and plants that left these indelible traces must have been minuscule and soft, constituted of single cells that liquefied soon after their death. This is why there remain no tangible imprints of their lives except for specific molecules more "noble" than those of bacteria: frail vestiges of the dawn of animal and plant life!

The time of their appearance coincides with a noticeable increase in traces of oxidized minerals.[11] These oxides indicate that the upper water layer was gradually becoming substantially oxygenated by plant-like bacteria. This fact supports the estimated date of the appearance of the first animal and plant cells, because only an oxygenated environment could have allowed their genesis.

The first fossil remains of large prototypical plants are astounding. The oldest date to 2.1 billion years ago. They were shaped like spiral tubes and were large enough to be easily visible (about the size of a piece of thin spaghetti) with no segments (they are not cells in a line) (see fig. 16d). Paleontologists are perplexed by these forms, all very similar to each other, that persisted on earth for a billion years, only to disappear completely.[12] Were they lines of bacterial cells with very fragile internal walls that did not fossilize? Or a unicellular alga containing many nuclei, just as several hundred species do today?

The other types to arise were minuscule organisms, different from all known bacteria. These petrified cells have a diameter (0.02 to 0.2 millimeters) more than twice that of the largest bacteria (fig. 15). A thick wall and a nucleus are sometimes visible (fig. 15e). The oldest date to 2.5 billion years ago.[13] Many similar fossils have been discovered throughout most of the world in rocks formed during the 1.5 billion years following their extinction.[14]

The first multicellular plants are seen in rocks 1.2 billion years old (fig. 16c).[15] But these rare fossilized algae are found only in several widely dispersed sites.

No fossil of an animal-like organism (engulfing organic matter or prey) consisting of one cell has been found in rocks dating from this period (between 2.7 and 0.85 billion years ago). Fossils of heterotrophic protists protected by a thick capsule were found in rocks between 850 and 742 million years old; they are classified among the modern testate amoebas (fig. 15h).[16] The dating of the first multicellular animals is also enigmatic. Only the holes or trails of burrowing, worm-like animals have been dated back beyond 2 to 1.8 billion years.[17] But other authors have contested the animal origin of these imprints.[18]

It is only beginning 570 million years ago that representatives of flora and fauna became significantly more common. Four successive layers with different faunas have been identified throughout the world in rocks that are essentially between 570 and 520 million years old.[19] This sudden

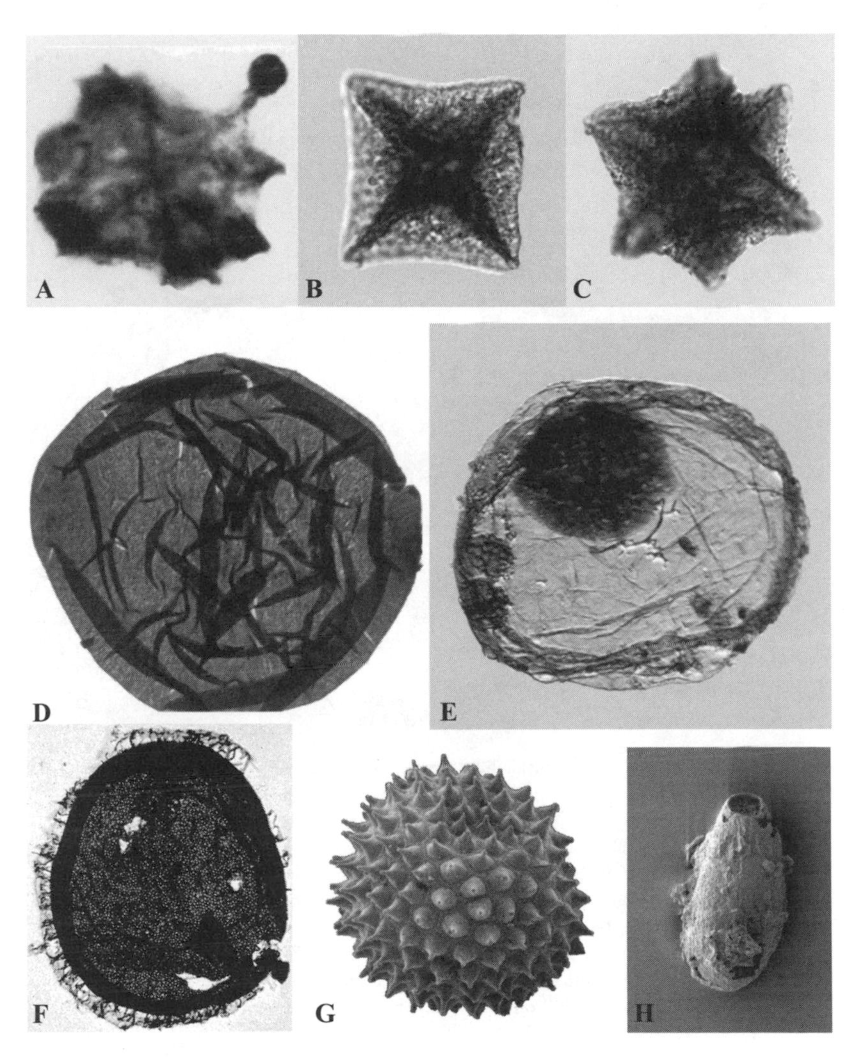

Figure 15. Astonishing forms of large cells, probably first fossils of microscopic unicellular plants or animals. *A*, Remains of an umbrella-shaped cell dated to 1.9 billion years ago, 0.02 millimeters in diameter. (Photo courtesy of J. William Schopf.) *B–G*, These large cells (between 0.02 and 0.2 millimeters in diameter) have been named acritarchs. They share a stiff cell wall similar to that of pollen grains. (*B–E*, courtesy of J. William Schopf; *F* and *G*, courtesy of Shuhai Xiao.) *B* and *C* are remains of angular cellular forms with a diameter (0.02 to 0.03 millimeters) similar to that of eukaryotes. *D* and *E* are large cells (0.17 millimeters in diameter) with smooth cell walls. *F* and *G* have cell walls bristling with expansions, which is common among acritarchs. *H*, Microfossil of a unicellular animal (0.1 millimeter tall) dated from 742 million years ago. This vestige of a capsule in the form of a vase originally contained a cell of amoeboid shape. (Photo courtesy of Susannah Porter.)

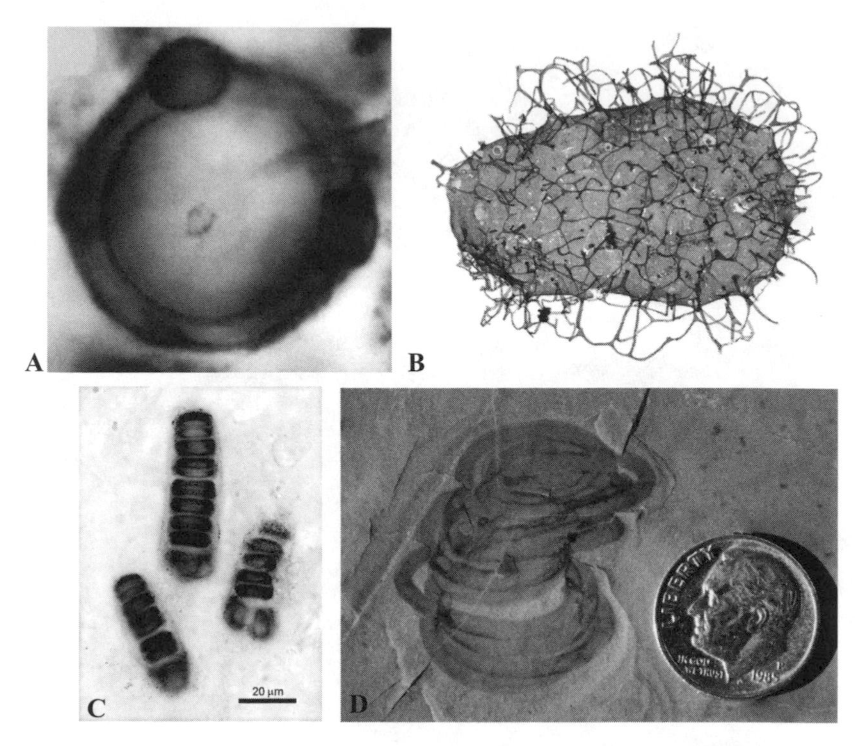

Figure 16. The first eukaryotic multicellular organisms, and those visible to the naked eye. *A*, The most primitive multicellular form, two billion years old. The large vesicle (0.02 millimeters long) is surrounded by a fringe of cells. (Photo courtesy of J. William Schopf.) *B*, Probable fungi (eumycete) dating from 800–900 million years ago. The elongated peripheral filaments of the large vesicle (0.3 millimeters long) are multicellular. The fusion of these filaments is a phenomenon best known in present-day "higher fungi." (Photo courtesy of Nicolas J. Butterfield.) *C*, These filaments of multicellular algae date from 1,200 million years ago. They closely resemble present-day red algae of the genus *Bangia*. (Photo courtesy of Nicolas J. Butterfield.) *D*, These filaments without partitions, visible to the naked eye, are found in rocks dating from 2.1 billion years ago. They are the oldest fossil filamentous algae. (Photo courtesy of Bruce Runnegar).

multiplication of life forms visible to the naked eye corresponds to the explosion of life marking the beginning of the Paleozoic geological era and the Cambrian period.[20]

The Emergence of Organisms Visible to the Unaided Eye

This brief history uses only fossil vestiges and their dates. One conclusion dominates: even if traces of animal and plant life visible to the naked eye are formally recognized in rocks dating from 2.1 to 0.57 billion years ago, the emergence of visible multicellular forms seems to have been the fruit of rapid evolution, as if some force led to their birth all at the same time, a few tens of millions of years before 570 million years ago.

Two hypotheses explain this sudden creative explosion of visible life.[21]

1. There was an earlier appearance of multicellular life that did not leave traces.
2. Some sudden impetus allowed multicellular life to flourish.

To justify the hypothesis of an earlier appearance of multicellular life, we would have to demonstrate that the many fossils dating from 570 million years ago descended from a genesis of much older lineages. To this end, it has been observed that the oldest petrified forms usually share a particular characteristic. Aside from several rare fossil deposits of soft organisms that left black carbon imprints,[22] the fossilized species have been preserved only thanks to solid substances that they constructed while they were alive. These are either protective walls of resistant organic matter (similar to that of the pollen grains of flowering plants, our nails, or the carapaces of lobsters) or calcareous structures (algae that fix calcium are still very numerous; among animals, mollusk shells, sea urchin tests, and coral skeletons are well-known examples). The accumulation of these solid substances appeared late in the evolution of multicellular flora and fauna (570 million years ago). Their preservation in sediments became possible then. We can deduce that before they acquired these structures, during the long span from 2.7 to 0.57 billion years ago, the great majority of multicellular organisms were soft and unprotected. After they died, their organic matter dissolved in water without leaving a trace.

Moreover, during this entire period, Earth's movements were substantial. The muddy shrouds of these organisms were often disturbed, recast, eroded, buried, heated, and compressed, erasing the rare, delicate imprints that would otherwise have been discovered today.

In considering on the one hand the truly poor fossilization of soft prototypical fauna and flora and on the other hand the effect of the Earth's movement in blurring traces of life, it is unsurprising that we can detect but an infinitesimal part of the original animal and plant biodiversity. Fossils leave us only a biased image of the real living diversity of the distant past. Therefore, we can understand that the fossils of multicellular algae from more than a billion years ago are the rarely found evidence of the precocious evolution of plants towards a complex state.

To justify the second hypothesis of a truly sudden, late appearance of multicellular forms, most researchers rely on a crucial ecological characteristic that could have caused life to blossom. According to them, the concentration of oxygen did not reach a level compatible with multicellular life until 600 million years ago.[23] To accept this hypothesis, we must know that in an oxygen-poor environment there is a real biological advantage in remaining unicellular and microscopic. This environmental limitation on the size of organisms derives from a simple geometric relationship. To absorb oxygen better (as well as salts and carbon dioxide), plant cells must develop a surface area in contact with the water that suffices to nourish the volume of the organism.

Evolution would then have been constrained by a simple geometric law: the larger the volume of any geometric solid (sphere, cylinder, cube, or parallelepiped), the smaller the ratio of surface area to volume. For instance, for a cube the volume grows as a function of the cube of a side, while the surface area grows only as a function of the square of a side. It is easy to visualize this property. Pour the contents of a bottle of wine (or soda or water), which represents living matter, into little glasses, filling them to the brim, and compare the total surface area of the glasses and of the liquid touching the air with that of the bottle. You will easily see that the surface area of the glasses and the liquid touching air is much greater than the surface area of the bottle, while the volume of liquid is exactly the same. The ratio of external surface area to volume is much greater with the glasses.[24]

This property is fundamental for any viable development of independent cells whose only mode of nutrition is absorption through their external walls (as is true for all unicellular algae). They must always have an elevated ratio of the surface of the walls (which are permeable to exchange of substances with the exterior) to the interior living volume. Beyond some limit of too large a living volume, the ratio becomes a handicap: there is no longer enough permeable surface to absorb enough inorganic food and gas for the interior material to live. This threshold corresponds to a microscopic volume.

Thus the diameter of the great majority of unicellular planktonic organisms, both fossil and present-day, does not exceed a tenth of a millimeter. In water with a low concentration of nutritional minerals, as is presently the case in the crystalline waters of the tropical Pacific Ocean or the Mediterranean Sea, a unicellular organism must be smaller still in order to survive. This is the reason the size of certain modern planktonic plant cells is the same as that of bacteria; they are part of the nanoplankton (between 0.05 and 0.005 millimeters in diameter) or picoplankton (less than 0.005 millimeters in diameter).

Between 2.7 and 0.57 billion years ago, in the atmosphere and the upper layers of the oceans, there was only between 1 percent and 10 percent of the current oxygen concentration. For plants in these conditions, to organize in colonies would have been a bad strategy, because such a form would reduce the ratio of surface area to volume. For instance, with two joined cells, the surface area in contact between the two cells is no longer in contact with the water, and this fact causes a deficit in surface exchange of materials with the exterior, so less oxygen will be available.

The absence of any unicellular animal remains during the entire period between 2.7 and 0.85 billion years ago can be explained differently. But first we must remember that the study of present-day cell contents shows that all unicellular representatives of the fauna and flora arose by addition of bacteria (by endosymbiosis; this genesis was explored in chapter 4). Now, in the process of adding bacteria (leading to the prototypical animals and plants), one cell always dominates (the one that captures a prey item and is of an "animal" nature), and the other cell is dominated (the prey, often plant-like). This endosymbiotic mechanism, today universally accepted by the scientific community, allows us to deduce that prototypical unicellular animals existed before the appearance of plant prototypes.

To chase and swallow living or inert microscopic prey, these prototypical unicellular animals must have been soft-bodied, like the great majority of those that still exist today (they comprise part of an ensemble of organisms called protozoans that arose in several different genealogical lineages). Their mode of ingesting prey by simply surrounding the cell, then dissolving it with digestive liquids, dictates that they should be roughly of a size able to swallow their minuscule prey.[25] The traits of these microscopic hunters of soft, mobile, unicellular prey would not have been favorable for fossilization.

All these arguments suggest that the presence of very ancient remains of unicellular algae (well fossilized and increasingly numerous, dating from 1.87 billion years ago and later) implies the concomitant presence of unicellular animals (though fossils older than 0.85 billion years have yet to be found).

In short, in the conditions of life in the sea between 2.7 and 0.57 billion years ago, to be large or to live in colonies would have been a handicap. Any transformation in this direction would likely have led to elimination by natural selection. During the first five-sixths of animal and plant life on our planet, to maintain a size similar to that of the first microscopic prototypes would have been the best strategy.

Even today, oxygen-poor aquatic habitats are inhabited only by unicellular organisms, the only ones that can tolerate this hostile environment.

To explain the highly localized discoveries of multicellular algae dating from 1.8 to 0.57 billion years ago and the simultaneous appearance of so many multicellular species at the beginning of the Cambrian (570 million years ago), some authors have hypothesized the existence of an inland sea (or several?), richer in oxygen, in which biodiversity would have had more time to evolve. At the beginning of the Cambrian, the oceans would finally have been well oxygenated and would have been inoculated with life by the opening of inland seas, oases of multicellular animal and plant communities that had long been evolving.

These theories are all interesting; they should be borne in mind and tested as evidence becomes available. Reality probably corresponds to some intermediate hypothesis. It is certain that the many multicellular animals whose remains have been found in several deposits dated from 570 million years ago and later were not elaborated all at once. They descend from lineages of multicellular species that began with small, soft-bodied forms. Their evolution surely required a substantial amount of time. They needed tens or hundreds of millions of years to attain a size barely visible to the naked eye. These archaic animals were gradually transformed to become better suited to a sedentary existence. They diversified and dispersed throughout our planet before their remains could, by chance, be conserved and survive intact until we discovered them.[26]

The development of multicellular organisms must therefore have begun well before the sudden appearance of remains of a size visible to our eyes (beginning 0.57 billion years ago).

The analysis and interpretation of genetic data from the most primitive modern animal lineages is disconcerting. It demonstrates the uncertainty that use of this method alone imposes on the ancient history of evolution. In fact, according to some authors, the divergence of multicellular animal lineages (and therefore their appearance) dates from 1.2 to 1.0 billion years ago.[27] This claim has been countered by other authors who estimate that the diversification of animals visible to the naked eye dates from 670 million years ago.[28]

Ecological Hypotheses on the Best Strategies to Live in Water in the Past

At this point, I will return briefly to a human history that greatly helped me to construct these histories of the geneses. I want to recall the life of my favorite painter: Vermeer. It happens that his life was quite mysterious. Few writings from his lifetime document his existence. And, above all, he left few works. Experts have attributed to him just 33 paintings, of which only three are signed. All his biographers have struggled to understand why his production was limited. They sought motives or circumstances that might explain this fact.

- Was he so wealthy that he did not need to be highly productive?
- He had indeed married the daughter of a landowner who had several properties, but he died drowning in debt and his large family (he had 11 children) had to fight to keep one of his masterpieces (*The Art of Painting*).
- Was he so much of an outsider that art dealers were uninterested in his works?
- We know that he exchanged several of his canvases to cover debts. For instance, an art dealer who came to Delft to see Vermeer's paintings found none in his workshop; he had to go to Vermeer's baker to admire one of the artist's works. However, other testimony compares him in his lifetime to Carel Fabritius (1622–1654), another Dutch master who was as well known in his time as Rembrandt. In fact, Vermeer excelled in his art; he was recognized and integrated into the community of artists and dealers in his city and was a master and member of the council of Saint Luke's Guild.

Vermeer's meager productivity will always remain a mystery. While it seems that circumstances favored his reputation and everything was in place to stimulate his productivity, he painted only two or three paintings per year. In 1672, the war and the annexation of the Dutch provinces by Louis XIV precipitated his decline. He died penniless at only 42 years of age.

All our questions relate to his means of subsisting, his behavior, his strategy, his success, or his decline. These questions fit remarkably with those raised by the appearance of multicellular organisms. In effect, paleontologists and biologists also seek to decipher ways of living. Thus, the questions on strategies of survival and on behavior allow us to understand the equilibria of life in a given environment. In the case I have been discussing, we must understand the reasons a diversity of organisms adopted another lifestyle when everything seemed to favor remaining small and unicellular. Why did some species become large and complex? These questions are in order because, whatever the timing of the appearance of visible life and whatever the oxygen concentration, to remain microscopic has always been a good strategy to live in the sea.

In order to reveal the causes of this change in status, we must add to the paleontological data those that deal with the biology and environmental conditions of organisms that lived on Earth between 2.7 and 0.57 billion years ago.

First, we should remember that life existed only in water. On land there was nothing: everything was inorganic. It was not until 470 million years ago that the first plants and animals began to conquer land by colonizing humid areas.[29]

Before the appearance of visible life forms, aquatic life consisted of a mixture of bacteria and prototypical unicellular plants and animals. Light energy allowed organisms with chlorophyll to live. Prototypical animals ate plants. At first, they were minute. The domain of this microcosm consisted of a single, clearly delimited habitat: the surface layer of the ocean, the only zone that was both sunlit and well oxygenated.

However, it is important to distinguish two habitats: the layer of surface water on the open seas between 0 and 100 meters deep, which today covers almost two-thirds of the planet, and the thin, linear volumes constituting the layer of water above the smaller coastal floor above the continental shelves, between 0 and 50 meters deep. Two modes of life correspond to these two situations:

- In all of the well-lit surface layers of the open oceans, the lives of marine organisms are always nomadic, vagabond.
- Above the continental shelves, essentially between 0 and 50 meters deep, sedentary or static life histories were able to evolve.

In the fluid environment of the open seas, where organisms are unattached to the seafloor, the food chain is composed of a lower level that develops because of the light. These are planktonic plants (phytoplankton are plant-like bacteria and microscopic unicellular plants that evolved where light could penetrate, as deep as 100 meters). The top level is composed of minuscule animals called zooplankton that eat plant cells.

In the narrow bands of water paralleling the coasts, the only oxygenated seafloor that gets enough light energy for photosynthesis, the most primitive multicellular organisms (basically masses of cells) can persist. Sedentary or static organisms therefore evolved. This mode of life has been adopted by:

- unicellular plant-like organisms (bacteria or prototypical algae) attached to the seafloor.

- multicellular species attached to the seafloor, like the plant-like bacteria (they have left many fossils, called stromatolites) or several algal lineages (which have left rare vestiges from before 570 million years ago).

The first multicellular animals, found in deposits mostly dating from 570 million years ago, are predominantly related to anemones, jellyfish, corals, or sponges. They all capture organic matter suspended in the water. At first they were very small and could sustain themselves by eating unicellular planktonic species. They were mostly attached to the bottom, so they had to cobble together, by progressive evolution, systems to capture the microscopic prey that drifted against their walls. These consisted of a suction and filter system for the sponges and tentacles for members of the large coral family.[30] Other species are worm-shaped; these were the first scavengers. They could sustain themselves on organic detritus and cadavers of plankton that accumulated on the seafloor, forming a layer of matter that could be assimilated. Thus, all these primitive animals, whatever their modes of nutrition, needed a regular supply of unicellular organisms.

This is a logical deduction: in order for multicellular life to evolve on coastal seafloor, there had to be not only oxygen but also dense, well-established "pastures" of unicellular species. It was therefore the accumulation of organic matter in a well-lit, oxygenated layer of water that led evolution in the direction of multicellular animals.

We will now try to tally the advantages and disadvantages of the conquest of these coastal habitats, limited parts of the marine domain but the main workshops for the evolution of prototypes of multicellular life.

No one knows if at the dawn of animal and plant life there were the same proportions of oceans and land as exist now. No one knows where the continents were then. Some geologists think that, at the beginning of life, dry land was abundant and dispersed throughout the oceans like lumps in a soup. In this view, these islands would be merged, then broken apart several times, and then rejoined. Our tangible evidence goes back only 1.1 billion years when all land constituted one big mass, a continent called Rodinia.[31] This continent then broke apart 700 million years ago, and the parts scattered as new, smaller continents. These continents regrouped 270 million years ago, forming a new supercontinent called Pangaea. Pangaea fragmented in turn, and its pieces drifted to form the current configuration of continents.[32]

It has also been established that at certain times land and oceans were almost entirely frozen (this was the "snowball Earth").[33] Also, various meteorites have slammed into the Earth, causing disruptions as great as those that led to the disappearance of dinosaurs 65 million years ago.

Therefore, it is highly likely that, between 2.7 and 0.57 billion years ago, the climate of Earth was very dynamic, with cataclysmic crises (geological, meteorological, volcanic, or caused by meteorites) that were more frequent and larger than those we know better that have occurred during the last 570 million years. In these conditions, the nomadic habits of microscopic algae and their minuscule unicellular animal predators would have been advantageous. In the vastness of the oceans, there is always a zone in which the slow or violent environmental changes could still be survived. Microscopic planktonic cells were able to persist in sanctuaries—masses of water in tropical seas always remained at least temperate during the coldest periods. Plants and animals that had embarked on a sedentary or static mode of life (that is, having a stage in the life cycle that was lived on or near the seafloor) were more exposed to cataclysms. Their migration towards less affected regions was more difficult than it was for the floating planktonic cells.

The first minuscule multicellular animals, which consisted of several tens or hundreds of cells, thus arose, possibly in successive waves, in coastal areas blessed with long periods without cataclysms. In their primitive states, they could not move of their own volition in water. It is important not to think of the first animals as having fins, flippers, or similar means of propulsion. They were surely tiny, flaccid, and with reduced mobility (for cells with microscopic cilia) or immobile. It was therefore on or attached to the seafloor, in weightless conditions, that the evolutionary handiwork of the multicellular animals passed the most rigorous tests.

Aside from these considerations of the environmental causes of the discrete and late emergence of multicellular life, we must not forget the main characteristic of the flowering of animal and plant life during this period (between 2.7 and 0.57 billion years ago). *Life was essentially planktonic,* composed of microscopic, nomadic individual cells. Only a high density of this living soup allowed both a perceptible increase in oxygen concentration and the emergence of the first visible animals.

The Demands of Life in Open Water

A behavioral trait was shared by all these organisms living in open water in the surface layers of the seas and oceans. All microscopic planktonic species needed both oxygen and light. These two conditions were and are still today found together only in the surface layers of water, basically in the first tens of meters of depth. These organisms, therefore, had to stay in this vital layer of water. Now, unicellular organisms and the primitive colonies of more or less aggregated cells cannot swim in such a way as to remain in this layer. Either they must develop rough surfaces in order to sink slowly and gently and be occasionally carried or transported towards the surface by currents, or they have to have the same density as water, if they are to stay where they have to for their entire lives. Therefore, they must regulate their density to remain at one depth, always in the surface layer.

How would I be able to give an example of this ability?

A new word struck me as helpful. Just as *submarine* means "under the water," *subfloating* can mean "floating below the surface."

I was fairly satisfied with this term but not convinced that the reader would understand the originality of this trait. So I will use a simple, more common example. While writing these lines, I was vacationing in Corsica as a festival of regional foods was taking place in the small town of Luri, on the Corsican Cape. Most exhibitors were vintners. It was a festival of wine (the Fera di u Vinu), very well known throughout the island. Visitors were invited to taste thirty high-quality vintages from throughout Corsica. My attention was drawn to an isolated exhibition that sold various tools needed to make wine. It was there that I found my inspiration; among the gadgets offered were a number of densimeters. A densimeter consists of transparent glass vial with little weights, surmounted by a long, thin, graduated glass rod. One immerses the densimeter in a liquid to determine the density of that liquid compared to that of water. The weighted vial subfloats, so to speak, with only the end of the glass rod emerging above the surface. The density of the liquid can then be read on the graduated scale on the rod; one looks for the marking right at the surface. This measurement tool was displayed with a name new to me: mustimeter. The exhibitor explained to me that it is used in the first phase of making wine, at the time of the initial crushing of the grapes, when the must (the grape seeds and skins) is still mixed with the juice. The mustimeter allows the vintner to measure the density of the juice before it ferments; this measurement gives the concentration of sugar in the liquid, which foretells the alcohol concentration the wine will have. For each liquid, there is a densimeter: for wine or alcohol, it is called an alcoholometer, for seawater it is a salinometer (fig. 17a). Each type of densimeter is weighted according to its designated use (alcohol is less dense than water while seawater and the juice of must are denser). The weight is adjusted by the number of small lead balls sealed in the vial, which always subfloats. A toy, the Cartesian diver, operates on the same principle, but it is the pressure that makes the figure rise or sink. The Cartesian diver (often represented by a figurine suspended in a perforated sphere or plastic bottle) subfloats, floats, or swims, depending on the pressure one exerts on the water in which it is immersed. Similarly, another object can illustrate the principle of subflotation. This is a very creative decorative thermometer. It is made of a glass column containing a liquid and glass spheres of different weights that float, subfloat, or rest on the bottom (fig. 17b). Each sphere is marked with a number that indicates a temperature. As the temperature changes the water density, this thermometer gives the right temperature—the number on the sphere that subfloats.

Planktonic cells that subfloat are analogous to densimeters with variable weights. The cell is constantly changing its weight so that it subfloats, compensating for the endless changes in temperature and atmospheric pressure that affect the density of the water. Plankton that have solid walls of calcium or silica compensate for this weight by accumulating lipid droplets. Often, such walls are very irregular, conferring a roughness or rugosity that slows the inexorable tendency to sink. A finer degree of regulation is achieved by the continuous absorption or excretion of mineral salts. In some cells, such as plant-like bacteria, gas microbubbles called aerosomes form. These mechanisms of weight adjustment are all crucial for microscopic plankton species. The subtle and simple imperative of staying in the surface layer of water (often above thousands of meters of depth) is a major constraint on evolution by natural selection. New mutations that do not account for densimetric equilibrium are likely to be eliminated in the first generation. In effect, every species that does not respect this equilibrium sinks into the shadows and is deprived of light or floats to the surface where it is burned by ultraviolet rays. Consider what happens after these cells die. The continuous weight adjustment ceases, and the resulting equilibrium is always negative: the cells sink and end up contributing to the calcareous or siliceous marine sediments. Only multicellular organisms that have landed by good fortune on patches of near-shore seafloor have a chance to pursue evolution towards complexity.

Present-Day Descendants of Microscopic Subfloating Cells

Algae in present-day plankton, descendants of the original plankton, are always minuscule. Though they are still unicellular plants, constrained

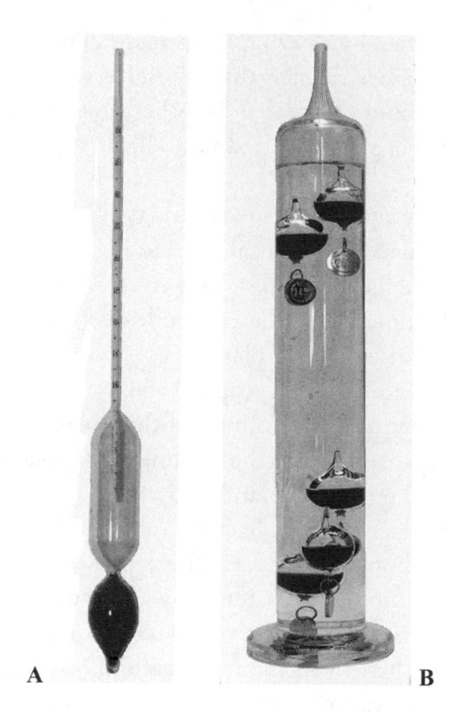

Figure 17. *A*, A densimeter. *B*, A thermometer with subfloating spheres with temperature medallions (also known as Galileo thermometer). (Photo by Alexandre Meinesz.)

by a highly restrictive set of requirements (always remaining small to assimilate enough gas and nutritive salts and to subfloat), they have been transformed over hundreds of millions of years and have continued to live and reproduce in the face of two other constraints: planktivorous animals and climatic changes. The result of contingent, random evolution is therefore oriented towards producing all the cellular equipment needed to maintain densimetric equilibrium, all the while improving defense against animals that would eat them and resistance to environmental hardships, such as cold temperatures or lack of nutriments.

Today, the number of unicellular planktonic algal species approaches 110,000,[34] while multicellular algae include just 9,000 species[35] (99 percent of these macroscopic algal species are confined to the narrow patch of seafloor along the coasts[36]). The plant biodiversity of aquatic habitats is therefore dominated by microscopic species. During more than two billion years of evolution, the walls of independent plant cells have been festooned with spines, carapaces, and shields of calcium or silica. These structures for defense or to slow sinking all aid the ability to subfloat by increasing friction, thereby slowing movement. These architectures and ornamentations are more diverse than those of the shells of all known marine mollusks. Toxins elaborated by other unicellular algae to render them less appetizing are more varied that all the molecules of our pharmacopoeia.

The weight of phytoplankton species today is still far greater than that of all multicellular algae and coastal marine higher plants combined. If we bear in mind that the seas and oceans cover 71 percent of the Earth, we understand why unicellular plants preside over most of the surface on which plant life has arisen on our planet.

Finally, if we consider that phytoplankton are spread throughout the upper 100 meters of water, the volume of this "plant water" (part of the hydrosphere) is much greater than the volume of the "plant air" (the atmospheric layer in which terrestrial plants grow).

As I developed the chapter on bacteria (chapter 3), I emphasized their preeminence throughout our planet. This led to an evident conclusion: evolution does not automatically tend toward complexity. This dictum also applies to plant life: it is not necessary to be complex and large to dominate and to profit best from the main natural resources of our planet. Two and half billion years of evolution of plant and animal life, marked by a multitude of transformations, lead to the same conclusion: it is the smallest, the minuscule forms composed of a single cell that have always been best adapted to life on our planet, and probably always will be.

These few words of Roman naturalist and philosopher Pliny the Elder (ca. 23–79 AD) apply superbly to this chapter:

> *Naturamusquam magis est tota quam in minimo.*
> Nature is never as great as in its smallest creatures.[37]

A century and a half ago, the theory of evolution by natural selection from the simplest to the most complex was placed in doubt by its author. Darwin did not understand the absence of any form of life before the sudden appearance of a great number of multicellular forms that were already diverse and highly evolved.[38] Today, "Darwin's dilemma"[39] is definitively resolved by solid evidence—discovery of the oldest fossils. But above all, this dilemma no longer exists because of the shattering of a fundamentally narcissistic dogma, which was a red herring, a false scenario for the evolution of life. To support this dogma, the only organisms considered were those corroborating a gradual ladder of increasing complexity of life leading to us. All species that had not embarked on that royal way were simply ignored. Seen as inferior beings or arrested stages of evolution, they were not included in the first schemas of evolution illustrating the march towards progress. They did not merit attention.

However, the microscopic, living Cartesian divers that comprised the earliest plankton today bear witness to the dominance and diversity of a different mode of life than that of complex organisms. They have evolved at their own pace, accumulating improvements that have adapted them to their environment, all the while remaining microbes. They do not constitute an evolutionary dead end; rather, they are representatives of major evolutionary pathways.

Many questions remain about the genesis of multicellular forms and pathways to increasing complexity of life. The metaphor Darwin cited is still germane: "the geological record [is] a history of the world imperfectly kept, and written in a changing dialect; of this history we possess the last volume alone, relating only to two or three countries. Of this volume, only here and there a short chapter has been preserved; and of each page, only here and there a few lines."[40]

The passage to multicellular life consisted of chapters and volumes that have disappeared. But the few lines recently discovered and new deductions have allowed us to read the archives anew.

CHAPTER 7

The Lego Game

The Genesis of Visible Life

From an early age, my son Jean-François was an avid Lego fan. More than any other of my four children, he was most likely to become engrossed in fitting together the colored bricks to make interesting shapes and various models. From the age of three, while at preschool, Jean-François began his Lego apprenticeship. The pieces for that age group were oversized so small children could not swallow them. In the pediatrician's waiting room, children could also play with Lego pieces, which helped them to forget their nervousness over the strange person who occupied those quarters and in whom their parents had such great faith. Even the children's shoe store had a play area provisioned with Lego blocks for antsy children, to distract them so their mothers could calmly select the right footwear. In this Lego environment, some children are really excited by the game, others are left cold. Jean-François was among the former, the Legophiles. This game is very educational: the child can look at the pieces, choose from among them, then assemble them into free-form sculptures or according to a set of instructions provided with the game.

I spent a lot of time, not playing Legos with Jean-François, but picking up and putting away his pieces when he finished. This was especially true during the learning period when he could not yet read the instructions.

Hundreds of unmatched pieces were contained in a big box . . . the heritage of his sisters and brother. Everything began by spreading them out: he wanted to view his treasure all at once. The hundreds of pieces would therefore be spilled all over the rug. He always marveled at the diverse colors, shapes, and sizes of the Lego bricks. For Jean-François, the apprenticeship began with a classifying game. I had to find him containers to help him group identical pieces into separate groups. Length, position of the locking pegs and holes—everything was identified and then classified and grouped according to such criteria.

One day, he was ready to put them together for the first time. The noise, the click when two pieces fit together, was an exciting discovery. At first he had fun creating simple shapes. He especially liked sticks made of stacked bricks. Once he had mastered that principle, he always either picked all bricks of one color or alternated two colors. Next he launched into making a wall or a floor. Parents: note that during this apprenticeship the act of assembly, however simple the form, demands great concentration. Frustration is quickly manifested by kicking piles of Lego pieces, hurling pieces all over the room. I am surely not the only father to have looked for scattered pieces under furniture.

Several months passed before he succeeded in constructing forms that resembled familiar objects: cars, houses. He was already a real initiate; he understood the creative power of Lego. Jean-François then advanced to the next level, where he could decipher diagrams of elementary Lego constructions. Simple models then began to grace his dresser. Then, much later, Jean-François became a master of the art of Lego assembly. The models were increasingly complex and . . . onerous. Thanks to his generous grandfather, Jean-François built one model after the other. One day, he called my attention to the fact that no real instructions were on the pages of the most complicated Lego architectural plans. One simply had to look at a page with a plan and see how it had changed from the previous page: the model was modular.

Today, Jean-François is nineteen years old, and he still keeps as souvenirs a great diversity of sophisticated Lego models in a glass display case: an aircraft carrier, many types of robots, planes, trains, and cars. Living entities do not lend themselves to Lego modeling: life forms are too rounded (though he assembled mythic, angular dragons and placed them behind his mechanical masterpieces).

When I came to the point of describing the genesis of multicellular animals and plants from unicellular organisms, I was inspired to use Legos as a metaphor: the bricks represent the cells, and the most elaborate models, made up of hundreds of different pieces, represent complex animals and plants.

For a biologist, this idea is preposterous. First of all, cells are so familiar to him or her that there is no need to describe them with metaphors. Further, the metaphor is peculiar because the elaboration of a living form is dynamic. The initial cell divides in two, the daughter cells then divide, and so forth. All the cells divide as an embryo grows into an individual. But the Lego bricks are static; they do not divide. To build a Lego model is to assemble preformed pieces according to a preconceived plan. This is true, but I also grant great importance to the didactic aspect of metaphors, to their power to aid understanding of abstract concepts. Having found no really common thing that suggests the formation of an object by tiny elements that divide (only life is capable of this process), I have therefore decided to keep the Lego game. It seems to suit the image of the emergence of multicellular organisms, and it is exactly this transformation of life that I want to illustrate.

As in the preceding chapters, I am trying to arrange our knowledge of the history of nature in a hierarchy, choosing the steps that seem to me most important. For the very long, ancient period (between 2.7 and 0.57 billion years ago) during which animal and plant life slowly emerged, there was only one fundamental step: passage from a solitary life to life in a community. That is to say, the transition between the single, independent animal or plant cell (the term unicellular refers to this state, still microscopic) and organisms comprising several joined cells (the term multicellular refers to this macroscopic state, usually visible to the naked eye).

Although this last complication of life concerns just a few animal and plant lineages, we should consider this change to be one of the three main steps in the history of life (with the two I have already described: the bacterial origin of life on Earth and the union of bacteria to produce unicellular flora and fauna).

It is a fundamental step, because if we take into account the multiple phases in the evolution of life, the collective passage (observed in distinct animal and plant lineages) from independent unicellular life to multicellular community life is much more abrupt and influential than the evolutionary changes between an alga and a moss, a moss and a fern, a fern and a flowering plant, or the transformation from a fish to an amphibian, a reptile to a bird, or a reptile to a mammal.

The union of cells is the initial evolutionary step that led to the diversity of organisms visible to our eye. It marks a beginning, the *genesis of complex life*.

The passage from microbial unicellular animal and plant life to multicellular life is easy to imagine; it seems so evident that is not discussed in most evolution books. Nevertheless, it required a series of major innovations, all of which had to be acquired by multicellular organisms.

I am first going to present the environmental causes that oriented life towards complexity, towards multicellularity. Then I will describe the various stages of cellular association that led to the emergence of multicellularity, and I will emphasize that they were not all predetermined, and there is no tendency towards complexity of life.

Finally, I am going to stress the very strange parallelism of multicellularity: the fact that diverse, independent lineages of life all adopted multicellular structure and specialization of cells. Is this coincidence a simple convergence of adaptation to a shared lifestyle? Or does it derive from a shared genetic basis that drives evolution in this direction?

First Step: Becoming Sedentary

I explained in the last chapter that, at the beginning of microbial animal and plant life, the ocean was the only habitat suitable for life. To subfloat in open water and to remain microscopic, nomadic, and independent is the best survival strategy. As soon as the oxygen concentration reached a level that allowed colonies of cells, colony size became strictly limited. Above a certain volume, a colony could not maintain itself in a state of subflotation nor remain in the vital space of the upper surface layers of water solely by means of the elementary propulsion systems of its members—minuscule whips called flagella (like the flagellum of a spermatozoon). This is why, in open water, the great majority of innovative evolutionary steps that led to the formation of multicellular colonies (at first they had to be minuscule) must have been fated to sink and disappear into the shadows of the great, oxygen-poor depths.

All animals swimming today in open water derive from organisms that evolved on narrow coastal seafloors (or even on dry land, as whales and porpoises did). The ocean habitat is at once homogeneous and demanding. On the whole, it is not propitious for the emergence of organisms remarkable for their size and weight. For multicellular species denser than water, the narrow oasis—oxygenated and well lit—of the narrow seafloor along the coasts is a favorable prospective habitat. However, it presents a new challenge. To live there, they must find food and also attach themselves to rocks or bury themselves in mud in order to remain in place and not be carried away by the force of waves, swells, and currents.

The oldest vestiges of multicellular life confirm this logical scenario. Algae, corals, sponges, and worms are the first multicellular organisms encountered in paleontological archives. They all lived side by side along the coast. All indications are that the unknown prototypes of these lineages, at first surely composed of several cells stuck together, were failures at planktonic life. By chance, they drifted down and landed on the continental shelves rather than the deep ocean floor. These shallow areas near the coasts were the only sites where conditions favored survival and the emergence of the first macroscopic organisms.

Therefore, in order for the prototypes of multicellular animals and plants to grow and evolve, two behavioral innovations were needed: to cease subflotation and to stay on the narrow floor of the continental shelves.

Look at the Earth from an airplane. Fly over the coast, the border between water and land. You see great hydrological and geological diversity there: coasts assailed by storms, swells, and waves; beaches uncovered and covered again by tides; calm bays; open or closed lagoons; lagoons with fresh, brackish, or salt water; estuaries; deltas; sandy or rocky shores; linear or jagged coastlines; cliffs; scree; warm or cold water, depending on the season or the region.

At the frontiers of these two substances—water and land—conditions of life change quickly. Erosion, accumulation of sediment, powerful currents that change abruptly, rivers that dry up or overflow, volcanoes that, in a flash, change the coast, building islands, which subsequently sink into the sea.

It is in these varied and changing seascapes, this palette so diverse and compartmentalized, composed of habitats that are often hostile, that the true cradle of diversity and macroscopic life is located. Many environmental constraints have favored the isolation of distinctive evolutionary paths; they have stimulated the diversification of multicellular life.

The littoral environment is hostile to the minuscule and fragile attempts at multicellular life, but it is diverse, compartmentalized, and changing, so it is favorable to rapid evolution of structural innovations. Many scenarios have been sketched out to reconstruct the details of what happened. The data allowing us to imagine how the transition occurred consist only of similarities with the strategies adopted by the simplest current living organisms.

The means of constructing present-day primitive multicellular organisms can be analogous to those used in the distant past. These are simply resemblances and need not constitute the real initial mechanisms. Or they could be homologous. In this case, the means of constructing the simplest current multicellular organisms are identical to those initiated in the past. Whichever the case, the most elementary strategies of remaining sedentary seen today furnish examples of successful attempts at a multicellular life.

I am a phycologist, a person who studies algae (from the Greek *phycos,* meaning alga, and *logos,* meaning study), a term preferable to algologist (from the Greek *algos,* meaning pain, and *logos,* meaning study) ever since departments dealing with the study of pain (in French, *algologie*) began to appear in French hospitals. Through studying algae, I have become familiar with their diversity. Though it is possible to identify many algal species from their shapes, sizes, and colors, we must often use microscopes to study their anatomy and sometimes their cells to refine our identifications. A thousand subtleties let us distinguish among algae that, at first glance, seem identical.

I remember long hours spent discovering algae of all types under the microscope and meditating on their astounding beauty and their simplicity or complexity. Thanks to this practice, I was led to examine many strategies leading to complex living architectures.

Everything begins with a preparation: an algal fragment in a drop of seawater between microscope slide and cover glass. Having framed the preparation on the microscope stage, I assumed the observation position, back straight and eyes at the oculars. First I had to look at the specimen under low magnification, centering it on the stage by pushing the slide lightly. I then had to change the objective, "going to 60." To do that, I simply turned a revolving barrel with many objectives and lined up in the axis of vision the one with lenses that magnify 60 times. The objective magnifies 60 times, the ocular, 10 times. We multiply to know the total enlargement: everything I was looking at was magnified 600 times. I also had to adjust the light intensity, with the aid of a rheostat and a diaphragm, to obtain a brightness suitable for observation. The first image was often fuzzy; to see the alga with clear outlines and details, I had to adjust the distance between the preparation and the objective (we do exactly the same thing when we look at something under a magnifying glass). To this end, I moved the microscope tube vertically in its housing, almost imperceptibly, thanks to a toothed wheel.

Everything was finally sharp and bright, and it was always a magnificent revelation! The colors and forms of microscopic life are as dazzling and complex as those of species we can see with the naked eye. I should add that I especially appreciate this minuscule world because I have learned to recognize, understand, and interpret the bizarre microscopic beings whose anatomy is revealed before my eyes.

Stroll in a tropical forest with a botanist. Where the casual spectator will appreciate the forms, foliage, and colors of the flowers, the botanist marvels at the diversity of species and is excited to recognize so many plants that he or she has seen in books and of which he knows the classification and biology. To appreciate nature to the fullest, it is necessary to know it.

Therefore, every time I look at algae, I am astonished by the richness of the levels of organization. Everything is different among the representatives of the plants that constitute very distinct evolutionary lineages (having arisen from various combinations of bacteria, as I explained in chapter 4).

To the green chlorophyll that they all possess are added, according to the particular lineages, brown, yellow, red, or blue pigments. Certain unicellular algae are equipped with one, two, three, or four flagella. Sometimes they have excrescences in their walls resembling shields or are protected by siliceous or calcareous carapaces.

Among the forms made up of assemblages of cells, the simplest are filaments composed of a dozen or so cells piled up end to end or forming balls or crusts. The most complex algae have diverse levels of cellular specialization and incredible sizes (like the brown alga *Macrocystis* that I had occasion to admire in the northern Pacific and that can be more than 60 meters long).[1]

As is evident, the study of marine plants, still living in the habitats in which multicellular life first appeared, develops the imagination needed to reconstruct the first unknown stages of evolution of the flora and fauna. It is doubtless in the diverse lineages of present-day algae that biologists can observe the strategies of elementary adaptations to a sedentary life. It is also possible to describe diverse important steps that led to multicellular life, without the risk of designing a model that is too specific to explain the genesis of a particular lineage.

If we can imagine that everything began with the loss of the state of subflotation, we can picture a cell (or a group of cells) unable to maintain itself in open water, sinking to a part of the coastal seafloor that favored its survival (it was sufficiently well lit and had enough oxygen and mineral salts).

Jean-François at the beginning of his apprenticeship: he has just dumped out his box of Lego pieces, resting on a table. The different pieces (colors and shapes) are jumbled on the rug. Jean-François likes to mix them, to spread them all over the rug.

The Lego bricks scattered on the ground (fig. 18a) are analogous to the diversity of isolated cells that sank to the coastal seafloor before the genesis of multicellular life.

To remain in the coastal fringe, yet not be forced to undergo dangerous abrasion and buffeting on the marine floor by the violent motion of the sea, an organism must bury itself or stick to a fixed part of the floor (a rock rather than sand or mud, continuously moved by the sea). Therefore all the fortuitous mutations fostering this sedentary lifestyle were selected

Figure 18. Four Lego configurations: *A*, scattered; *B*, piled up; *C*, attached to similar pieces; and *D*, attached to different pieces. (Photos by Alexandre Meinesz.)

for by nature; they gave the initial advantages in living on the narrow coastal seafloor.

This way of life is still very common today. Many unicellular algae and even single-celled animals have lived this way for hundreds of millions of years.[2] Since they first appeared, these microbes have adopted and preserved this sedentary behavior without ever having tested life as a community of cells.

Second Step: Initiation of Social Life

First consider a scenario in which adopting a life in common with other cells is provisional and reversible. The first actors must have been represented by aggregations of identical cells. Attachments among them must initially have been weak and the place of an individual cell in the colony must have been indeterminate. These configurations of poorly organized cellular associations could have represented transitional phases leading to a stronger social cohesion with cells joined more firmly.

This type of assemblage, in which the number of cells and their organization vary from one colony to another, still exists today (fig. 19).[3] Some of these colonies can even subfloat in the form of balls or wreaths, but beyond a critical length and mass they are found only attached to rocks.[4] The association between gregarious cells of these contemporary colonies is still reversible. Each cell can still live independently.

Jean-François ends this phase of his apprenticeship: he has noticed amidst the jumble of pieces that some have identical shapes and colors. As a game, he selects them and places them in piles. Piles of identical pieces.

It is this result and not Jean-François's action that presents the analogy. It is a configuration of a colony of cells (fig. 18b). But these piles of identical cells have been formed by successive divisions of a cell. Each time, the daughter cells have remained close to one another to form a pile, a colony.

In some contemporary representatives of this type of colony, a gel (secreted by the cells) serves to bind several identical cells. Moreover, synchronous movements or similar behaviors, useful to the community, can be detected in the cells making up some present-day configurations of algae that live colonially (fig. 19d).[5] These could not occur without systems of communicating between all elements of the colony. This is the premise of a social, collective life in which coordination among all cells becomes necessary.

The Genesis of Multicellularity

The promiscuity of cells living in a colony and communicating among themselves can lead to the evolution of a conjunction of elements constituting a colony. The cells will then be attached to one another in a way that is irreversible for them and their descendants, because this trait will be inscribed in their genetic information.

Jean-François just after his discovery of the essential principle of the Lego game: joining the pieces together. His earliest, most elementary assemblages consisted of identical bricks attached to one another (fig. 18c).

Another strategy leading to this archaic stage of a collective life can also begin by ordinary cell division. The scenario is very easy to imagine. An

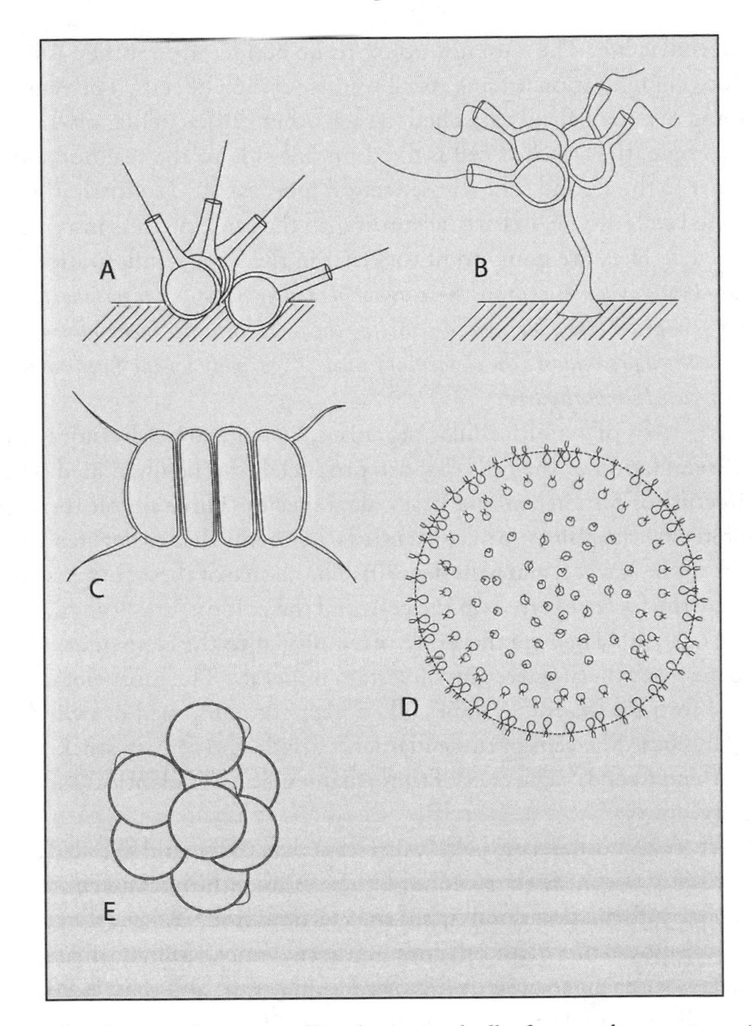

Figure 19. Different configurations of loosely associated cells of present-day organisms. *A*, Fixed (benthic) cluster of cells of a green alga (*Chaetosphaeridium*, Chlorophyta). *B*, Fixed (benthic) cluster of animal cells: flagellate protozoa, the closest relatives of animals, ancestor of sponges (choanoflagellates). *C–E*, Colonies of subfloating (planktonic) green algae (Chlorophyta) (*C*, *Scenedesmus*; *D*, *Volvox*; *E*, *Coelastrum*). (Drawings by Alexandre Meinesz, arranged by Marjorie Meinesz.)

initial cell, which has lost the ability to subfloat, is integrated into a coastal life and is attached to the seafloor, and it divides in two (this is simple vegetative reproduction, which produces a clone of identical cells).

The new cells no longer float as the ancestral cell had. Thus, the two daughter cells have no tendency to separate, unlike other representatives of unicellular life. They are not going to lie beside one another as in the previous configuration leading to a loose association of cells. The two cells are going to remain firmly attached to each other in a particular orientation (for example, the attached cell is fixed on one side to the seafloor and on the other to the second cell, whose weight it bears). And so forth: the two associated cells divide, in turn, according to the same process inscribed in their genes. They are going to fit together in the same configuration.[6]

Jean-François advances to another stage of his apprenticeship. He no longer enjoys grouping the pieces into piles before assembling them. He attaches them immediately. He has learned precociously the elementary mode of assembly for the Lego game and has constructed simple figures.

This genesis of a multicellular organism, by vegetative divisions without dissemination of the cells that are produced, can be illustrated by the architecture of certain contemporary algal species. Three simple forms are elaborated by the piling up of cells: crusts, filaments that either stand erect or stick to the surface, and balls (fig. 20). The choice of these forms is influenced by the forces exerted on the cells and the colony (gravity, viscosity, surface tension). They are therefore often similar to the elementary forms that determine the architecture of certain minerals. The most elementary of these forms is that of an unbranched filament composed of cylindrical cells (fig. 20d). The cells of these filaments divide and remain stuck to one another end to end.[7] The cells of these filaments are all identical and form a society.

But there is another, completely different way that could have led to the appearance of organisms large enough to be visible: the nucleus containing the genetic information divides, but the cell does not.[8] What arises then is a form composed of a giant cell containing two nuclei at first, then many. Nowadays some algae present this astonishing trait, and they can also be crusts, balls, or simple or branched filaments. This evolutionary pathway can even lead to the elaboration of complex forms resembling terrestrial plants that can reach several meters in length (although they are made up of only a "tube," with no walls, no compartments, no cellular structure). This configuration, a giant cell containing many nuclei, is also seen in some fungi[9] and even in lineages of animals that are essentially microscopic.[10] An intermediate stage is found in some other algae arranged in segments that resemble cells but that each contain many nuclei.[11]

Paleontologists were shocked to observe that this very unusual structure is that of the first petrified vestiges of organisms of visible size: these are algae in the genus *Grypania* (see fig. 16d in chapter 6)—fossilized in rocks formed beginning 2.1 billion years ago.[12] They have no internal cell walls, no cellular subdivisions.

To explain the emergence of multicellularity in visible life, some biologists have turned to these ancient and contemporary architectures without cells. They have imagined a scenario of delayed cell division.[13] According to these authors, giant cells containing many nuclei were at the origin of multicellular life visible to the naked eye, owing to a segmentation of the interior space that occurred *at a later time* (fig. 20a). A box (or cell) would, in this scenario, have been created around each nucleus inside the giant cell.

This route to development of multicellular forms by subdivision of a giant unicellular matrix does not correspond at all to a game of Legos. It resembles more closely the technical work that precedes the elaboration of a Lego construction. In effect, builders of a Lego model always begin with a form in mind (an automobile or a house, for example) then subdivide that mental image into Lego pieces. The Danish inventors of Legos were originally cabinetmakers. By dividing a figurative piece of wood into parts that could then be fit together, they invented Legos.

This hypothesis of the formation of multicellular organisms by division of a giant unicellular one with several nuclei was especially attractive as an explanation for the origin of flatworms (from a worm-like prototypical animal composed of a giant cell with multiple nuclei), but it was disproved by comparative analyses of DNA.[14] Therefore, at present, there is no evidence that this pathway to the evolution of multicellular organisms was ever taken. Biologists nevertheless remain perplexed in the face of the remarkable complexity of some species visible to our eyes but made up of a single cell in which certain parts play particular roles (some have the shape and function of leaves, roots, or branches, but internally, they have the structure of a simple tube).[15]

However, this was not the evolutionary trajectory taken by all animals visible to the naked eye, or the mosses, ferns, and flowering plants, or the great majority of algae and macroscopic fungi, because they each consist of an assemblage of thousands to millions of microscopic cells.

Some of the oldest fossilized multicellular organisms (they are eukaryotes and had one nucleus in each cell) date from 1.2 billion years ago. These are filamentous algae composed of lines of identical cells (see fig. 16c in chapter 6).[16] Even today, many species have this architecture (fig. 20e). Filiform algae, consisting of straight or branching lines of cells, are found among six of the nine genealogical lineages of chlorophyll-containing plants.

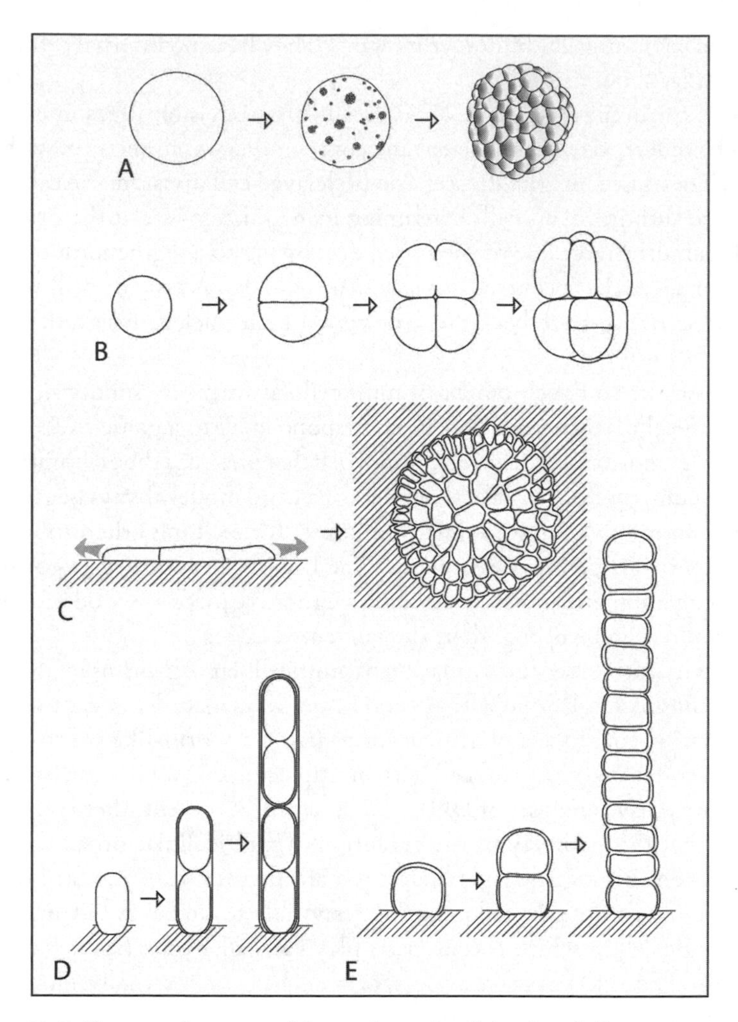

Figure 20. Different configurations of the simplest multicellular algae: balls, crusts, or filaments. *A*, Formation of a multicellular ball by internal compartmentalization and division in an initially multinucleate giant cell of a green alga. *B*, Formation of a multicellular ball by successive divisions in a planktonic green alga. *C*, Benthic crustose red alga comprising a single layer of cells forming a crust. *D* and *E*, Benthic filamentous multicellular alga in which each cell can divide; they are undifferentiated. *D*, In the green alga *Cylindrocapsa* (Chlorophyta), after each division, the two daughter cells remain attached end to end inside the original cell wall; a filament is thereby produced that is constructed according to the nested matryoshka doll model. *E*, In the red alga *Bangia* (Rhodophyta), a filamentous multicellular alga formed of several stacked cells. (Drawings by Alexandre Meinesz, arranged by Marjorie Meinesz.)

Among the many modern algae that have this architecture or a similar one, we can distinguish some in which all cells retain the ability to divide in two.[17] At each division, the line of cells elongates or thickens. Sexual reproduction often marks the death of this type of alga, because all the cells of the filament transform into sex cells that are liberated into the water.[18]

The sex cells then meet in open water, fuse, sink, and attach to a favorable substrate. They will constitute the origin of a new filament having the same architecture as the parents. The filamentous architecture is inscribed in the genes. The program of modular development of the filament (cell after cell) is also memorized genetically.

Division of Labor in a Society of Ancestral Cells

In other contemporary filamentous algae, we observe the first cellular specialization: only the cell located at the tip of the filament can divide.[19] The other cells are kept in check: this is the simplest manifestation of cellular differentiation. It is the beginning of the partitioning of tasks among associated cells. This method allows the association of cells to live better sedentarily or statically. The civilization of cells by assignment of different roles is now organized.

Another manifestation of cellular differentiation concerns the production of sex cells. It appears when certain cells—also called germ cells—are able to become fertile. The other cells, which always remain sterile, are called somatic cells.

The third step in cellular differentiation is imposed by innovations leading to a complex vegetative apparatus. To form architectures more elaborate than a simple filament, either a sphere or a crust, cells adopt various roles within the assemblage according to their locations in three-dimensional space. The sizes, shapes, and contents of cells with distinct roles become different (fig. 18d).

During the development of a complex organism, the processes of cellular differentiation follow subtle rules of transmission and programming of genetic information. The genetic information, though identical in all the cells of a multicellular organism, is used and transmitted differently according to the location of each cell in the community and according to the stage of development of the community.

How activation or inactivation of certain crucial genetic information is achieved is still little known. However, this mechanism is the basis of the partitioning of cellular forms and functions that allows a society of many cells to become a multicellular organism. A possibly homologous phenomenon determines the forms and functions of individuals in the most social animal societies, as is exemplified with termites and ants; within the anthill, all ants have neither the same appearance nor the same role.[20]

A communal, collective, social life is thereby elaborated.[21] Each multicellular individual is a structured civilization of cells. The result, the architectures of life formed from cells of different sorts, can be compared to Lego models composed of varied forms and colors.

Constructions consisting of several thousand Lego bricks of more than a hundred different types can be seen in Jean-François's display case.

For animals, the first multicellular configuration was neither a filament nor a crust, but a hollow ball[22]—a ball made up of a single layer of cells in contact with the external environment and with an internal, enclosed space (fig. 21f–k). It is striking that this initial configuration, a hollow ball, was also adopted (fig. 21a–e) and preserved in different present-day algal lineages (fig. 20b).[23]

The animal balls were doubtless soft, and their cells, all identical, must have caught a maximum of inert or living organic matter with their flagellae and transmitted the dissolved nutritive substances to the neighboring cells. This hypothesis is supported by some of the oldest animal fossils known today. They are shaped like balls composed of a layer of contiguous cells. Dating from 600 to 550 million years ago, they have been interpreted as sponge embryos (fig. 21f–k).[24]

Development of the Embryo: Retracing Ancestral Evolution?

The German naturalist Ernst Haeckel (1834–1919), inventor of the term "ecology," also proposed a theory of the emergence of multicellular animals from a colony of identical cells in the form of a hollow ball. He imagined such colonies to be the first stages of animal life in his drawings. His theoretical drawings have an uncanny resemblance to the oldest animal fossils known at the end of the twentieth century.[25] In fact, Haeckel also founded another theory, termed *recapitulation,* associating the evolution of animals (phylogeny) and the development of their modern embryos (ontogeny).[26] He observed that the early developmental stages of very different animals (such as those of an insect, a sea urchin, a frog, and a mammal) are very similar. The initial cell, formed from the fusion of a spermatozoon and an egg, divides according to the same configuration. The

Figure 21. Stages of development of multicellular organisms dated from 580 million years ago. *A–E*, Stages of development of algae. *F–K*, Animal embryos. (Photos courtesy of Shuhai Xiao.)

earliest stage, termed the *morula,* looks like a tiny mulberry or raspberry.[27] As in fruits consisting of little spherical elements stuck together, a cavity forms in the center. The next stage, the *blastula,* therefore resembles a hollow ball made up of a single cell layer.[28] Haeckel's theory stipulates that the various embryonic stages of contemporary animals represent the memory of evolution. Therefore, in his view, the first animals must have resembled in their mature state microscopic mulberries. The forms of the common ancestors of all multicellular animals are, according to this theory, recapitulated during the elaboration of the embryo in all animal species.

Haeckel forced biologists to notice the resemblance among a young tadpole, the embryonic stage of a frog, and a fish (frogs and toads evolved from fishes). Similarly, he emphasized the resemblance between a caterpillar (the early ontogenetic stage of a butterfly) and a worm (insects evolved from worms).

This "recapitulation" is also seen in the development of certain plants. Among mosses, the first plants to have conquered humid terrestrial habitats, we find a seemingly atavistic trait. In effect, the first developmental stages resemble a filamentous alga: the initial cell created by the germination of a spore first forms a branched filament composed of cells lined up end to end.[29] The recapitulation theory led Haeckel to push these comparisons too far. A heated scientific debate ensued. Today researchers agree that this theory cannot be applied uniformly to all developmental stages of all species.[30]

However, we can imagine what the initial stage of animal evolution was like thanks to certain minuscule animal cells we can see today in seawater. These are independent cells very similar to those of sponges, the animals with the simplest anatomy. Some representatives of this group of unicellular animals (the choanoflagellates) can mass and form colonies (fig. 20b).[31] Identical assemblages could have established themselves in the distant past and led to the sponges. This is a well-supported hypothesis, because on the one hand sponges are static, simple animal structures that are among the first fossil animals found, and on the other hand the sequencing of the DNA of sponges and contemporary choanoflagellates allows us to observe that these groups have many genetic homologies.[32]

Architectural Blueprints Endlessly Tested and Refined: Life Becomes Complex

It took Jean-François many unfortunate attempts before he succeeded in constructing his first interesting forms. His memory had to be good enough to retain the successful elementary assemblages in order to make them again the next day. He required well-communicated information in order to program a complex construction (represented by the plan provided in the Lego box). Selection among attempts, memory, and programming were added by stages and allowed the construction of a large Lego model.

The same processes occur in the elaboration of organisms made up of many cells. It took more than a billion years to pass from the stage of one cell to the one represented by us, by way of the important innovation in the stage consisting of two attached identical cells. Mutations and genetic mixing created changes that would not have been expected to occur randomly. The architectures adopted by different lineages of plants and animals were oriented by physical and chemical constraints that imposed certain forms.[33] Perfect or imperfect forms were selected by other environmental constraints (competition, better adaptation to food or reproduction). The changes that conferred improved survival or increased reproduction were perpetuated, then improved upon. In its genetic code, each species retained in its memory the ancestral organization plan and its improvements. Evolution was therefore modular.[34] Each stage is a functional step that can be improved upon (this characteristic is not present in a Lego model, where the intermediate stages of construction need not resemble anything).

The more complex the plan becomes, the more memory must increase. This is why genetic information determining structure is cumulative and divided among several chromosomes.

It is thus for all animal and plant species visible to the naked eye. We are an example of a complex organism composed of hundreds of millions of agglomerated cells comprising more than 200 distinct types (bone, liver, intestine, neurons, ova, skin, muscle, as well as cells that have retaken a degree of freedom of movement in the cellular society, such as white blood cells and swimming spermatozoa). However, all these different cells contain the same genetic information, but this information is expressed or suppressed according to the type of cell and its position with respect to others in the course of development.[35] A very organized system of molecular communication determines the order of appearance, the placement, and the type of specialization of cells as the embryo develops (this is ontogeny).

It is logical to note that many similarities observed between the architecture of organisms and different developmental stages of embryos of different organisms are actually often genetic homologies. I can cite, for example, the morphological and anatomical traits common to mammals and the great resemblance of the embryonic stages of different species up through a very advanced stage of development: 99 percent of our proteins and nucleic acid sequences are identical to those of a chimpanzee.[36] Similarly, the number of nucleotide bases in the mouse genome is only 14 percent less than that of humans.[37]

Stephen J. Gould presented a logical demonstration of the inexorable increase in complexity of one part of living organisms.[38] It rests on the fact that the simplest organisms known are bacteria.[39] This "simple" state cannot evolve towards a still simpler state, so it should be considered an insurmountable wall for evolution. From this fact, it is logical to conclude that the evolution of life beginning with bacteria is of necessity constrained to two directions only:

- that limited to all the adaptations to unicellular life (evolution is internal, within the cellular factory).
- that leading to the union of cells to form complex multicellular organisms.

These two evolutionary directions are unlimited.

I have emphasized in chapters 3 and 6 how the great majority of organisms currently living on Earth have remained "simple" (unicellular), even though they have never stopped evolving internally. In the interior of independent unicellular organisms, the palette of possible variations is enormous on the physiological level as on the anatomical level, such as the number, shape, size, and particular aspects of all the cellular components.

Likewise, the possible architectural types composed of masses of cells are infinite.

The cumulative result of these two evolutionary pathways, consisting of internal tinkering plus massing of cells, constitutes present-day biodiversity. It shows that unicellular life remains dominant and that, among multicellular organisms, the smallest and least complex also dominate (among animals, species of insects and worms are much more numerous than vertebrate species). In the sea, the number of individual copepods (miniature crustaceans) exceeds that for all individuals of other pelagic multicellular organisms added together. Therefore, when we seek an overview of biodiversity, we see that there is no general tendency leading towards large size and complexity. Life is not oriented towards increasing complexity, nor is it fated to become ever more complex.

Complexity is but one of two directions open to evolution, and this direction has seen modest use by a small percentage of species.

The myth of systematic progress tied to evolution and mystical evocations of the vital spirit, predestined to push life towards complexity, are only narcissistic impressions. Stephen J. Gould's demonstration leads to the conclusion that increasing complexity is not the raison d'être of the history of life. There is no preconceived genetic program contained in bacteria that leads to the development of the "superior" and perfect entities that humans see in themselves. We represent a magnificent Lego construction elaborated without a predetermined plan beginning with the genesis of multicellular animal forms. We are not the realization of a preconceived model; we are simply one of the most complex of the millions of models produced by the contingent process of the agglomeration of cells.

Plate 1. During the past several centuries, the pictorial representation of the first humans has changed greatly! In any case, they always have a navel, the scar of an umbilical cord that links them to earlier ancestors. Left, Cranach the Elder, *Adam and Eve* (1533). (Photo by Jörg P. Anders. Berlin: Gemäldegalerie. By permission of Réunion des Musées Nationaux.) Right, John Holmes, sculpture of hominids at the American Museum of Natural History. (Photo courtesy of Julian Hockings.)

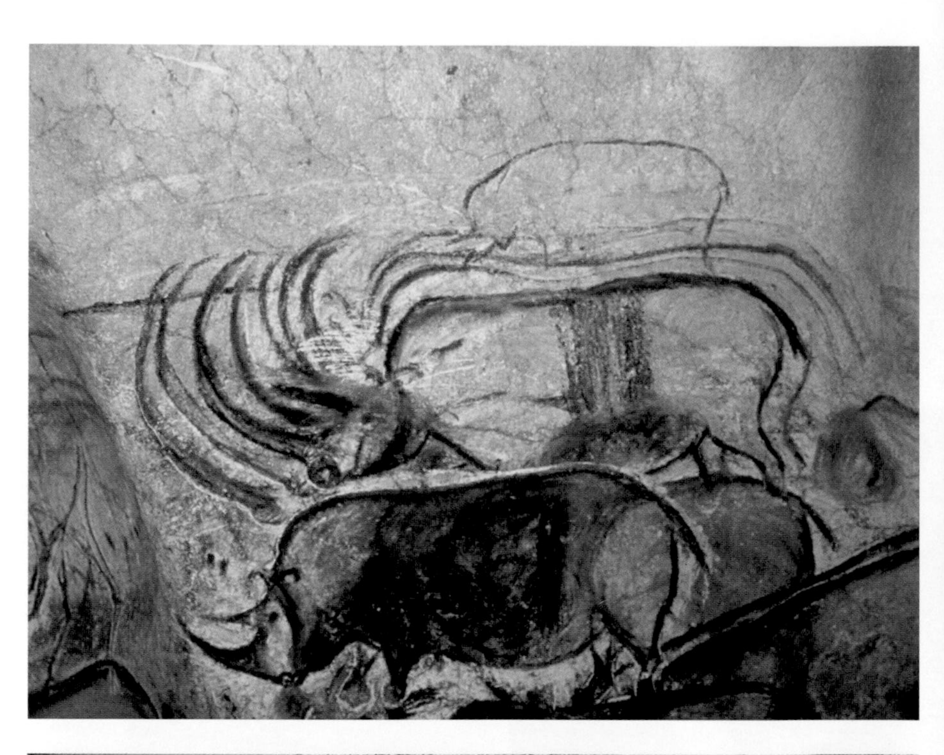

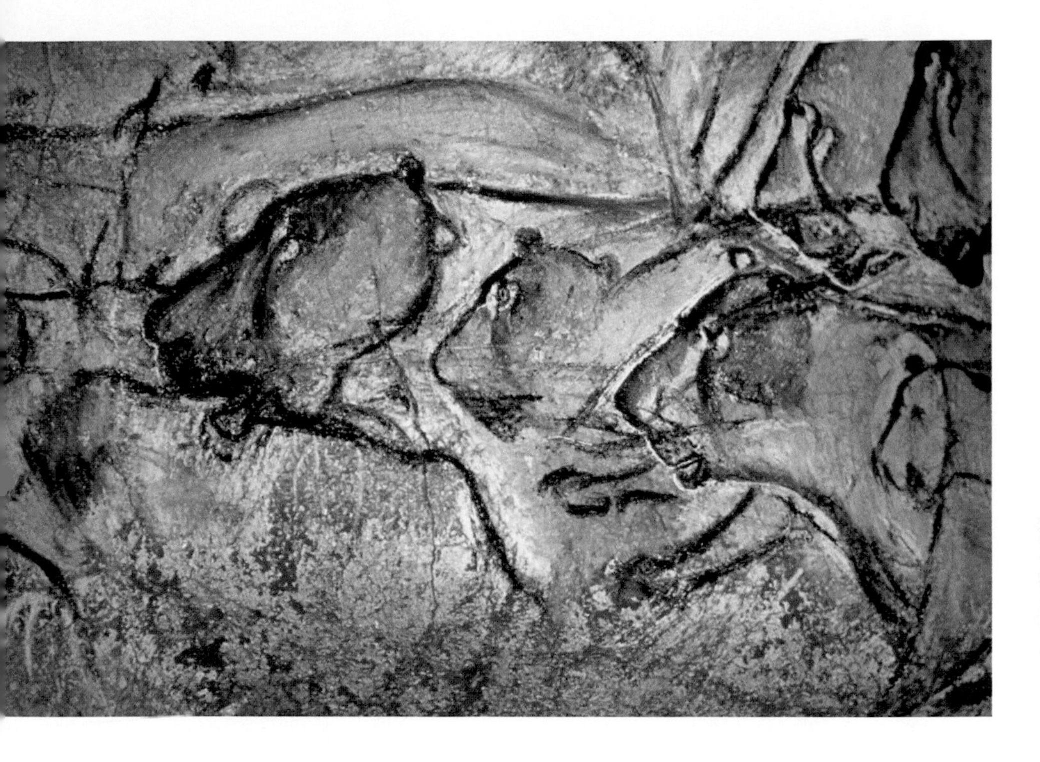

Plate 2. The first expressions of self-consciousness: the Chauvet Cave (30,000 BC). The first artists have recreated, realistically and with perspective, profiles of hordes of animals observed outside dark caves. I surely could not draw a group of cows (with no model, no less!) as well with charcoal on a wall lit only by flashlight. (Photos by Jean Clottes, courtesy of the French Ministry of Culture and Communication.)

A.

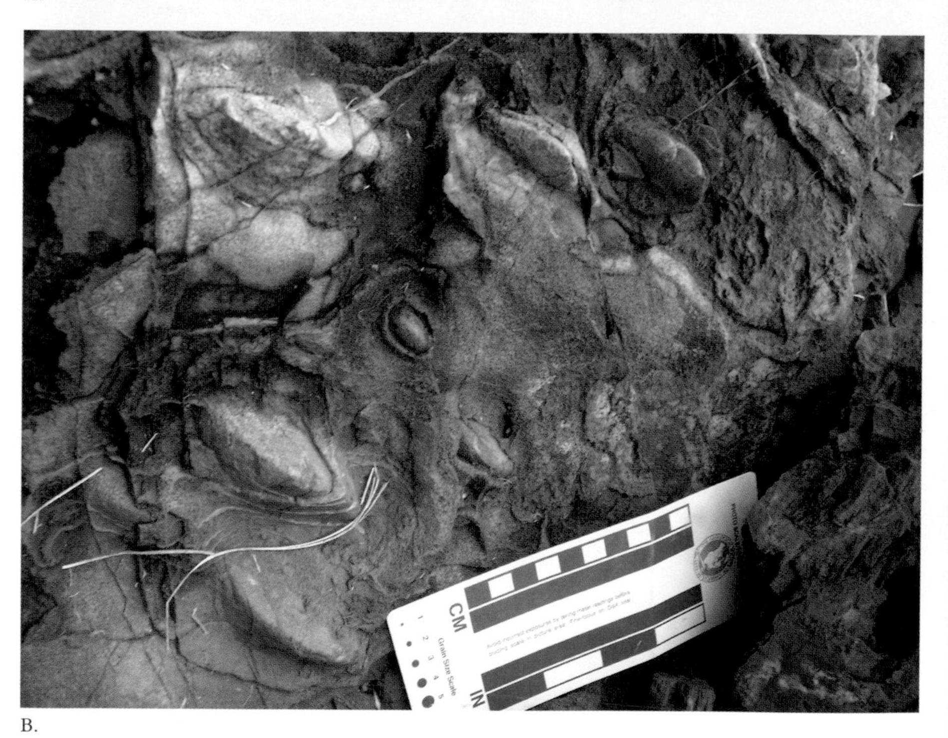

B.

Plate 3. The oldest Stromatolites. *A*, One of the oldest described stromatolites formed 3.45 billion years ago. The section of this rock shows the successive layers of sediment fixed in the past by filamentous bacteria. In this section we can distinguish two conical structures. (Photo courtesy of H. J. Hofmann.) *B*, Large complex cones, one of seven types found in a 3.43-billion-year-old stromatolite reef. (Photo courtesy of A. Allwood.)

A.

B.

Plate 4. Actual stromatolites. *A*, Present-day intertidal stromatolites. The upper part, in which are found living plant-like bacteria, is above the water line, which would have been impossible during the first three billion years during which life evolved. (Photo courtesy of Tourism Western Australia Image Library.) *B*, Present-day stromatolites developing entirely underwater. These stromatolites are in an ecological setting similar to that in which the first stromatolites developed. (Photo from Carribean Reef plant, courtesy of D. S. and M. M. Littler.)

Plate 5. The vampire slug on killer alga. These sea slugs have "invented" a sophisticated symbiosis. They retain the only useful parts of the ingested algae: the chloroplasts. They sequester the chloroplasts in intestinal diverticula near the surface of the skin. After eating, these sea slugs are green, and they activate their solar batteries, the undigested chloroplasts. From a vegetarian diet, the sea slugs switch to the plant nutritional mode, spreading their skin to capture the sun's rays. (Photo by A. Meinesz.)

Plate 6. Johannes Vermeer, *The Astronomer* (1668). Antoni van Leeuwenhoek, then age 36, the first person to discover microbes and their astounding diversity, was the artist's model. Louvre, Paris. (Photo by René-Gabriel Ojéda. By permission of Réunion des Musées Nationaux.)

A.

B.

Plate 7. *A*, Dr. Robert Paine on a bed of algae on Tatoosh Island during the summer of 2000. (Photo by A. Meinesz.) *B*, Dr. Hideki Ueshima showing his model of the Seto Inland Sea at the site of an especially heavily modified coastal region. (Photo courtesy of H. Ueshima.)

Plate 8*A*. Johannes Vermeer, *The Geographer* (1668–1669). Antoni van Lecuwenhock, then age 36, was the artist's model. (Städel Museum, Frankfurt, Germany. By permission of Artothek.)

Plate 8*B*. Pierre Brissaud, (1885–1964) painting of Antoni van Leeuwenhoek. In 1939, the French illustrator produced a realistic reconstruction of the visit by Queen Mary II of England (who is shown here looking at "animalcules" through a microscope) to Van Leeuwenhoek's laboratory for Abbott Pharmaceutical Laboratories. (By permission of Light, Inc., National Library of Medicine, and Abbott Pharmaceutical Laboratories.)

It is worth observing that this beautiful demonstration does not explain the reasons for the passage from one system to another that marks the genesis of multicellular life. Stephen J. Gould estimates that this passage is simply a matter of which pathway was open, a sort of logical transition in the sense that it was one possibility offered to evolution.

Multicellularity: An Innovation Common to Many Groups of Organisms

The metaphor of the Lego game, which allowed me to illustrate this natural history of the genesis of life visible to the naked eye, has the advantage of emphasizing the most important point. In the evolution of complex life, everything begins with a fundamental innovation—a process that allows cells to unite and then specialize (differentiate) as they live in a community. This is the initial phase of the multiple constructions of life visible to our eye.

The Lego game was patented in 1958, and the patent pertained to a single technical invention: the system by which the pieces stick together. The inventors aimed to label their invention "the power to create." This power to create was enhanced by providing pieces of different sizes, shapes, and colors. The tag line "the power to create" also applies to the greater game of the association of cells; their uniting and their differentiation allowed the genesis of multicellular organisms.

Once cells acquired the ability to attach to one another and to differentiate, nature played the role of a child with thousands of different Lego pieces available, imagining plans in stages, in modular fashion—sturdy frameworks in harmony with the exterior forces that were physical, chemical, and biological. Each successful plan served as a basis for infinite improvements, for other architectures. From that point on, the field of three-dimensional innovation was open.

The appearance of animals, fungi, and algae composed of several cells seemed to result from a universal logic, that of the old adage: *unity is strength*.[40] It is this union that marks the beginning of multicellular life, of life visible to the naked eye.

It is noteworthy that this strategy had already been used in the elaboration of life. In effect, I explained in chapter 4 how the first animal and plant cells were formed by the addition of bacteria. Endosymbiosis, the fusion of several bacteria, conferred strength on animal and plant life. It was the origin, the genesis of eukaryotes.

Union by the joining of several identical animal or plant cells opened the way to cellular societies with the potential to conquer new environments: the narrow, lit seafloor of the continental shelves, and later, dry land. Life was therefore able to leave liquid habitats and spread over land.

Now that I have established this history, one fact should be noted: multicellularity and cellular specialization were adopted at the same time by animals, fungi, and various plant lineages independently; that is, they did not have the same unicellular or multicellular ancestor. Essentially, the latest classifications of life suggest nine independent "plant" lineages (all organisms with chlorophyll). There exist today unicellular representatives in these nine lineages, and multicellular forms are present in six of them (including the brown and red algae and green plants).[41] For the other organisms—animals and fungi—multicellular forms exist in two of the twelve distinct lineages (one fungal lineage and the genealogical lineage consisting of all animals visible to the naked eye).[42]

Thus, there exist eight varied, independent lineages with multicellular representatives. The remarkable fact that all these lineages acquired similar complex social structures is surprising. How can we explain this collective and simultaneous propensity (in all these distinct lineages) to adopt multicellular life followed by cellular differentiation?

First of all, we should recall that the acquisition of multicellularity and cellular differentiation appears to have been induced as an effect of a specific cause: becoming sedentary. The genesis of multicellularity is therefore nothing but a logical consequence of adapting to survive in similar circumstances. We must then evaluate two hypotheses:

- either mutations, selected by nature to confer on organisms the multicellular state well adapted to life on the coastal seafloor, have appeared *independently* in all multicellular lineages;
- or all multicellular lineages had initially (in their unicellular state) the same information allowing them to become multicellular when conditions were favorable (when they became sedentary).

Multicellularity: A Product of Convergent Evolution?

If the first hypothesis is correct, multicellularity would have resulted from a straightforward opportunistic adaptation to prevailing environmental circumstances, adopted by different lineages that therefore converged evolutionarily. The term *convergence* refers to the adoption, by distinct lineages, of a similar optimal solution to a shared problem.

Physical forces or chemical laws act on molecules, cells, and the scheme of cell divisions and lead automatically to the contingent cobbling together of mutations and genetic mixing, after which the action of natural selection leads to the adoption of similar multicellular architecture, no matter the lineage (all lineages thus begin as filaments, balls, or crusts).[43] The emergence of the multicellular state would then be a simple self-organization in response to a shared cause.

This is a *materialistic* hypothesis of the genesis of multicellularity. Some authors prefer the term *systemic* to describe this system of formation of multicellular life, in order to stress that it cannot be explained solely by reducing an organism to its genetic information (as would be required by a fully *reductionist* logic).

In the history of life, examples of convergent strategies are legion. In particular circumstances, independent groups can adopt similar forms, sometimes eerily similar. One of the best known examples is of porpoises and sharks. Both of these marine animals have a fusiform body with fins that allow them to move swiftly in water. However, they belong to very distinct groups: porpoises are mammals, while sharks are fishes. They have converged; the bodies of these animals were progressively transformed by selection, which favored innovations that allowed rapid swimming. Similar constraints gave them similar shapes.

To illustrate convergent phenomena in biology, Stephen J. Gould cites the example of two extraordinarily similar characteristics that were evolved by two very distinct organisms: a mollusk closely related to clams and a type of angler fish.[44] With a part of its body, each animal has fashioned a lure resembling a small fish (of course this did not happen overnight; it took hundreds of thousands of years for evolution to achieve this result). The angler fish waves the lure in front of its enormous mouth and swallows the fishes drawn to it. The mollusk moves its fish-like lure to attract a fish; when the curious fish approaches, the mollusk ejects its larvae toward it. The larvae then live as parasites in the gills of the fish. This example is among the most striking cases of convergent evolution.

The paleontologist Simon Conway Morris ascribes a less anecdotal role to convergences. Citing many examples, he strives to show that convergences are evidence of a major evolutionary tendency. In his view, they show that there are a finite number of evolutionary pathways. They are canalized by many physical and chemical constraints.[45] For Conway Morris, the pathways available to evolution are therefore quite limited, and this limitation produces convergences. The Nobel laureate Christian de Duve came to the same conclusion; in his view, the possible outcomes of chance events are limited.[46] Evolution is therefore a selective optimization that consequently produces convergent forms.

Therefore, because convergent adaptations are so common in evolution, we can imagine that the origin of so many multicellular lineages results from an optimal strategy for a sedentary life. We must then consider increasing complexity as a convergent path that allows species to adapt to a coastal life. This view leads us to see an increased probability that the entire evolutionary process leading to multicellularity will be initiated during the adoption, for various reasons, of a sedentary lifestyle for unicellular organisms. Among the reasons, authors have suggested selection for better production of sex cells or the emergence of systems of cellular cooperation to find food better in a fixed position or buried in marine sediments. In either case, the process would be, in a sense, self-organized, whatever the initial genealogical lineage of the ancestral unicellular organism.

The Innovation of Multicellularity: Result of Shared Software?

The second hypothesis explaining the genesis of multicellularity points to genetic information shared among all unicellular ancestors of multicellular lineages. In this case, the organisms would have been preprogrammed to become multicellular when they became sedentary. This hypothesis posits a *deterministic* phenomenon (multicellularity would have been predetermined).

Just as for interpretations of evolution (chapter 5), various currents of thought produce different perspectives on this determinism. For many people, it is God alone who created complex multicellular life, with both a perfect work in his image—man—and all that is useful to man (the creatures of the fauna and flora). According to diverse religions, nothing has ever evolved; everything was preconceived, everything was set up in advance, everything was predetermined, everything was *designed with intelligence*. All life is therefore sacred. God is considered to be the "Great Architect of the Universe" (a phrase used in certain Freemason oaths), the "Great Conceiver of Life" (a phrase used by Jehovah's Witnesses), and, more recently, the "Intelligent Designer."

The Judeo-Christian and Muslim religions are based on this divine power of the conception of life and man. The entire story is explained well in the writings of the first prophets. Some Eastern religions are based on a planned conception of life entailing reincarnation of organisms that have lived previously.

This deterministic conception of life, consistent with religious doctrines, was cast into doubt beginning in the seventeenth century by several courageous philosophers who dared to formulate a divergent opinion. We should recall the Dutch philosopher Benedict (Baruch) Spinoza (1632–1677), who succeeded in expressing his ideas, then very controversial, while being careful to protect himself by defining God to be nature (*Deus sive Natura*). The two following citations best illustrate his revolutionary view: "*Natura finem nullum sibi praefixum habet*" (nature has no fixed predetermined goal) and "*Omnes causae finales nihil nisi humana sunt figmenta*" (all final causes are but human fictions).[47]

Spinoza wrote his principles between 1662 and 1677, only several kilometers from Vermeer's studio—while Vermeer was painting *The Astronomer* (in 1668)!

Of course, we can also construct another explanation for a shared determinism that impels unicellular organisms inevitably to become multicellular when they become sedentary. But on this point I have to state that the following reasoning is purely personal and that some biologists would be glad to call me a heretic wallowing in unfounded speculation. To elaborate this other scenario of the appearance of multicellularity, I see great importance in the appearance of two complex mechanisms that seem to me to be fundamental to the construction of multicellular societies:

- the *system of attachment and association between cells* (the fact that they divide but nevertheless remain firmly attached, then act jointly through communication systems)
- *cellular differentiation* (the fact that, during development, cells specialize and adopt different roles and forms)

These innovations seem exceptional to me, because they allow social life and the construction of complex forms. It is logical to suggest that the occurrence of events leading to such specific innovations has an infinitesimal probability of occurring in two distinct lineages.[48] But they are seen in eight distinct lineages!

The Program "Unity Is Strength"

To explain the determinism of multicellularity and the concomitant cellular specialization in many genealogically independent groups, I hypothesize specific genetic information or programming held by all ancestors of multicellular organisms, programming that was activated when these ancestors became sedentary. I am going to call this genetic information or programming, held in common by all these organisms (but not yet discovered by science), "unity is strength."

This mechanism held in common does not mean that animal and plant life was destined to become complex. In effect, it is on this point that I must link this hypothesis with three elements that the scientific community agrees on.

First, there exist plant-like bacteria that are highly structured, forming lines of cells, sometimes with branching, and that can produce greatly differentiated cells (not specialized in the sense of acquiring a specific form, but destined to take on a specific function such as capturing atmospheric nitrogen). These cyanobacteria, multicellular plant-like bacteria, were long called blue-green algae (or Cyanophyceae) (see fig. 5, chapter 3).[49] Now, some of the first vestiges of life, dating from 3.5 billion years ago, are multicellular bacteria similar to these plant-like bacteria. These ancient bacteria lived on the narrow continental shelf seafloor and were the basis of the stone carpets, the stromatolites (plates 3 and 4). They are evidence of the very precocious evolution of cells with specialized functions and of multicellular forms adapted to a sedentary life in this habitat.[50] The social builders (multicellular) of stromatolites were precursors of a system of sedentary life that may have given them a type of eternity: their very similar descendants, with unchanged habits, are still alive today.

Second, analyses of genomes of present-day bacteria and the study of their origin give evidence of a great mélange of information.[51] The genome of each bacterium is a patchwork of information coming from a multitude of exchanges of small amounts of genetic material. All groups of bacteria have inherited genes from or exchanged genes with other groups. Thus, the Archaebacteria (or Archaea, which I have called chemobacteria in chapter 3) and the Eubacteria (mostly plant-like bacteria and fungus-like bacteria) have genetic information in common.

And third, the genesis of unicellular animals and plants came from the fusion of bacteria (endosymbioses; see the discussion in chapter 4).

Therefore, the genetically based union of cells forming a civilization leading to "unity is strength" (multicellularity) appeared in ancestral bacteria at least 3.5 billion years ago (and cell differentiation appeared in plant-like bacteria beginning 2.45 billion years ago). As this is a matter of a heritable mechanism and is obligatorily passed from generation to generation, it is reasonable to imagine that it would have been transmitted later in certain bacteria that participated in endosymbioses that gave rise to animal and plant lineages beginning 2.7 billion years ago. Among the bacteria swallowed by bacterial predators and tamed within the bodies of the latter, giving rise to the first animal and plant cells, why not suppose that certain ones carried the genetic, social information of "unity is strength"?

I want to emphasize that the mechanism of multicellularity arose *three billion years* before the explosion of multicellular animal forms and that it has not been adopted by all the unicellular organisms that lived in the past or live today in a sedentary or attached manner. The great majority of modern bacteria and many unicellular plants and animals live on or in marine or terrestrial soil, yet have remained unicellular. The mechanism of convergence (the first hypothesis) leading by self-organization to multicellularity was not applied to this multitude of unicellular organisms in the course of several billion years in a situation favorable for its evolution.

My hypothesis explains this apparent anomaly: all those organisms that remained unicellular in sedentary situations simply did not inherit the genetic information for "unity is strength."

The mechanism that remains to be discovered would then be the analog of a method of using a computer that the users could improve upon indefinitely. When this system is activated, the organism that benefits from it can use its innovational principle (multicellular life with differentiated cells) to memorize information or programs that lead to successful assemblages. This hypothesis also leads us to consider that the forms of viable assemblages are perforce logical. Their coherence rests on physical and chemical forces that impinge on all the constituent parts of the model; self-organization would act only as the means of cellular association: forms and unions of molecules through masses of cells. Each step in the development of structure, well adapted to the environmental conditions, encoded in the genetic information, is like a personal subprogram for each lineage that is added to the basic model common to all.[52]

In this case, the initiation of multicellularity and the cellular differentiation of animals and plants are not purely convergent phenomena. The capacity to evolve in this way depends in the first place on the presence or absence of the ability to create complexity. It is thus determined by the presence of information that can be transmitted and that allows the innovation.

This is the major innovation of the Lego game. Before Legos were invented, children could only make ephemeral productions: they piled up pieces of wood of different shapes and colors until they collapsed (Jean-François also played this game, onomatopoetically called "Badaboom" after the sound of collapsing constructions). It is the system of sticking together that created the ability to make lasting forms.

The hypothesis of information held in common (so far just deduced, rather than demonstrated) is both reductionist and systemic.[53] It is a logical explanation for the concomitant acquisition of multicellularity and cellular differentiation in different lineages of animals and plants.

As we consider this hypothesis, we can ask ourselves about the origin of the transmissible social mechanism common to all these independent lineages. If we accept that life was created from inorganic matter and carboniferous molecules on Earth, then the mechanism of "unity is strength" must have arisen by evolution on Earth. If life first appeared elsewhere (see chapter 2), then the Earth would have been inoculated by bacteria carried by meteorites. Was the mechanism of "unity is strength" already present among these bacteria?

The two possibilities lead to the same result: bacteria (probably plant-like) dating from 3.5 billion years ago were already multicellular! Now, the probable period of the emergence of life on Earth before the date of the oldest fossils of multicellular life (3.5 billion years ago) must have been very short (relative to a geological time scale) with environmental conditions that were hostile to the blossoming of life. I therefore prefer the hypothesis that the information for "unity is strength" was already present in bacteria that arrived on Earth—they, or at least some among them, already had the capacity to organize in multicellular forms when this complex state became very favorable for life. If this speculation turns out to be correct, this propensity was initiated and must have been developed elsewhere. At this point, I find myself staring into an abyss of mysteries.

I do not want to prove the existence of little green Martians; rather, I want to show that a sequence of plausible hypotheses suggests that visible, well-structured life forms exist in places other than Earth. I leave to filmmakers or novelists the job of imagining what they might be like, and they have been doing just that for a long time. My reasoning simply aims to evaluate the likelihood of the existence elsewhere of multicellular forms with differentiated cell types. And if I am correct, the causes and the circumstances leading to the creation of the information that impelled life to become complex remain mysterious.

Here again I am aware of the cascade of suppositions and speculations underpinning my reasoning. But this is just an attempt. Scientific research will test this hypothesis, which seems plausible to me.

Architectural Blueprints and Division of Labor: The Third Genesis of Life

This personal representation of the greatest period of evolution of fauna and flora, which occurred between 2.7 and 0.57 billion years ago, can be summed up in three main points:

1. The development of complex life constitutes only one evolutionary pathway; it is not the general rule. Unicellular organisms (bacteria and single-celled animals and plants) still dominate life on our planet by their numbers and global mass. They have continued to evolve internally.
2. The behavioral trait of becoming sedentary and the constraints of the marine coastal environment led to a strategy of collective life. Societies of cells, well adapted to a sedentary lifestyle, emerged by the successive addition of cells.
3. Multicellularity and cellular specialization were acquired by very distinct animal, fungal, and (multiple) plant lineages. This innovative life strategy was adopted either through convergent paths of self-organization or because of the expression of information or programming that was held in common. In this latter case, all the complex, visible life forms have the same mechanism, which had the same origin, that allowed them to formulate organization plans for societies of cells, and these were improved from generation to generation. This mechanism, held in common, could have arisen and been expressed elsewhere.

For all these reasons, I have considered the acquisition of multicellularity and cellular differentiation as the emergence of a new potential. It marks the third genesis of life, that of the flora and fauna that are visible to the naked eye. Once these principles of the union of cells and cellular specialization were acquired, the contingencies of mutation, genetic mixing of sexual reproduction, and physical and chemical constraints (which impose various basic forms) created the living architectures that were then selected according to their abilities to feed themselves, defend themselves, and reproduce. This genesis of societies of cells corresponds to the acquisition of a three-dimensional creative power. Beginning with this genesis, the fantastic array of visible life colonized the seas, then land and air.

Biologists and colleagues: recall that the concept of the cell is abstract and that a cell is never seen in the course of a life of a non-biologist except for rare laboratory exercises in school. Remember also that the simple idea of affirming that an ape could have evolved into a human seemed so strange and inconceivable to our ancestors a mere two or three generations ago. Understand that the idea that our direct ancestor can be a bacterium whose oldest tomb dates from 3.5 billion years ago is even stranger.

Also, I thought that a bit of imagination could better demonstrate this conceptual gap. In this context, my reference to Jean-François's favorite game is appropriate, because all parents and children know the ingenious system by which blocks stick to each other. To analogize a Lego block to a cell is certainly too trivial and imprecise for biologists, but it enlightens the uninitiated.

I have led you to the era in which Antoni van Leeuwenhoek discovered cells swarming under his magnifying lens. I hope that with the Lego game I have helped you to understand that life experienced a creative thrust thanks to the mechanism of collectivism, the adoption of a social life organized among many cells. The conceptual gap and the abstraction might therefore be partly alleviated. Yes, it is necessary to see in every living organism a beautiful model of Lego pieces of different forms and functions. Yes, you can imagine the beginning of its construction by the assembly of elementary pieces. Yes, you can understand more complex assemblages by reference to the apprenticeship of Jean-François, who had to acquire sufficient memory and use a construction plan to bring his models to fruition. Yes, you can even understand the dilemma about two academic hypotheses by reference to the Lego metaphor:

- Must we attribute Jean-François's successful joining of Lego pieces simply to his stage of maturity, applied to all the types of Lego pieces spread out on the carpet within his reach?
- Or do we have to consider above all the pre-existence of a system of joining that allowed him to stick the pieces together?

The many shapes and colors of living organisms that we can admire today can be depicted in a giant painting as spilling out of a cornucopia.[54] But the elements of this painting, the biodiversity of Earth visible to the naked eye, are not fixed. Their stationary images on the canvas account only for the instant when they were painted. Everything keeps changing, nothing is fixed, all will change tomorrow. Living works are never finished.

Vermeer created paintings we admire today. My father was also a painter, and I remember his hesitation before the subjects of his paintings, the changes he imposed on the models, beginning with his first brushstrokes on the blank canvas. So I can well imagine how Vermeer must have taken his time to frame his models and give them lifelike, realistic postures. He also had to create the desired ambiance, in harmony with the persons he painted. The clothes, the decor, the lighting—everything was conceived before the first brushstroke. Then he refined the play of light, the facial expressions. He had to respect the geometric laws of perspective and proportions in sketching with a camera obscura and vanishing points on his canvas. In his day, pigments did not come in metallic tubes; he had to grind organic or inorganic substances and mix them with binders. This is how he created his paints. Everything was therefore thought out, preconceived, created by the artist. And finally, a painting, a work created for eternity. A canvas, but a fragile one, exposed to the ravages of time or to possible catastrophes like fire, theft, or attack by a madman. As it happens, Vermeer's *The Astronomer* barely missed disappearing forever. Once owned by Baron Alphonse de Rothschild, it was stolen by the Nazis during World War II. The work, confiscated by Rosenberg, was coveted by Hitler. Fortunately, after the war, the painting was found and restored to the Rothschild family. In 1983, France acquired *The Astronomer* as settlement of a debt; since then, it has been in the Louvre.

In the long genesis of multicellular life that has led to us, how many attempts or models were needed in order that, among all the architectural sketches of life arising fortuitously, a little wormlike organism emerged, progenitor of the vertebrates, of which we are one species? The little worm and its many cousins were not protected by the noble future of their lineage; they survived multiple planetary disasters by luck. Thus a second question is in order: *how many other lineages with promising futures disappeared?*

The brutal extinction of lineages shows again that nothing is definitively won or predestined in evolution. Everything can be threatened by an accident. Now, accidents on a global scale are many—these are the great cataclysms that have devastated our flora and fauna at various times. In the history of life, they should not be considered as secondary phenomena just because they are straightforward and do not require much intelligence to describe. They have determined the present composition of life.

CHAPTER 8

Candide, *Jurassic Park*, and Noah

Voltaire (1694–1778) was remarkable for his erudition and significant contributions to literature, philosophy, history, and geography. His writings ran up against a prevailing mindset derived notably from the religious doctrines that dominated at the time of King Louis XV, nearly a century after the wars of religion. A believer, but one who detested dogma, superstition, and intolerance, Voltaire was in permanent revolt and produced his critiques tirelessly, despite the hostility his writings engendered. *Candide, ou l'optimisme,* published in 1759, is his most famous tale.[1] Juxtaposing irony and excess, Voltaire attacked certain "politically correct" dogmas of his time.[2]

The story of Candide is just a long series of mishaps, ghastly calamities, and horrible catastrophes endured during an extended voyage through Europe, America, and the Caribbean, during which Candide constantly questions his worldview. Pangloss, his tutor, considers all the disasters Candide encounters to be "for the best." For him, everything that happens is always good because everything is programmed: "It can be demonstrated that things could not be other than they are: for everything has been made to serve a purpose, and so nothing is susceptible to improvement."[3]

A song, popular in France between the world wars, would have been a favorite tune of Pangloss; its refrain is, "Everything is fine / Madame the Marquise / Everything is fine / Everything's going great!"[4] The irony is that the verses relate catastrophic situations for the Marquise.

Pangloss, an extreme representative of the philosophical school of stoicism, constantly professing optimism, is convinced that every event that confronts man in his life is predestined and that we must accept them all because they are planned "for the best." By the effect of the excesses of accumulated hardships, horrors of war, injustices, and natural catastrophes experienced by Candide and his traveling companions, Voltaire ridicules this reasoning. He derides the religious doctrine of a divine providence and the variants of stoicism[5] that tend to justify the most horrendous fates of men or to view them as programmed, predestined.

Among the catastrophes depicted in *Candide,* the terrible earthquake that destroyed Lisbon, the capital of Portugal, stands out. This quake struck on Saturday, November 1, 1755, killing 60,000 souls, nearly a fourth of the population of the city. Many victims perished in the collapse of churches, where they were attending All Saints Day services. For Pangloss, the cataclysm was planned and "for the best." To the unfortunate survivors, he announced: "If there is a volcano under Lisbon, then it couldn't be anywhere else; for it is impossible that things could be placed anywhere except where they are. For all is well" (p. 12).

For Voltaire, the Lisbon earthquake was an example of an extremely consequential event, arbitrary, totally unexpected, not predestined, and against which the will of man was helpless. He refused to consider it as divine punishment or as an end programmed by nature. For him, this natural catastrophe was a model of woeful injustice. In June 1756, he made it the subject of a poem, the conclusion of which is summarized in these two lines:

> One day all will be well, that is our hope:
> All is well today, that is our illusion.

We can always delude ourselves, but history just happens; no divinity plans it.

The multiple disastrous, unexpected, life-altering events that seemed to follow Candide on his journey were used by Voltaire to call attention to the fragility of existence. Everyone has an uncertain, unforeseeable, sometimes unjust destiny, whatever his or her attitude and faith.

The tale of Candide's life can illustrate a fundamental characteristic of the pathways of life. In effect, the history of nature as a whole has been marked by a succession of apocalyptic cataclysms that have each devastated biodiversity. These cataclysms have led to a blind selection of certain lineages and species. If the ancient mammalian ancestors of our lineage were spared the injustice of a premature decapitation, it was because of blind luck.

Thus, the cataclysms represent a fourth layer of randomness, or contingencies, for the history of life. This layer is added to the three other layers of randomness that govern the appearance of mutations, the genetic mixing of sexual reproduction, and the environmental circumstances of existence. This fourth sort of randomness that intervenes in the evolution of life is better known nowadays thanks to a well-publicized example: the event that caused the total disappearance of dinosaurs. This catastrophe affected all life extant at the time, before the advent of man.

Jurassic Park

When the film *Jurassic Park* appeared, I stood in line to see it with my children. I saw it twice more, and I saw *The Lost World: Jurassic Park*, *Jurassic Park III*, and the animated cartoon *Dinosaurs*,[6] by Walt Disney Productions. I noted that the splendid reconstructions of these reptiles had an extraordinary impact on both children and adults.

In the *Jurassic Park* series, man encounters hostile dinosaurs resuscitated by entrepreneurial scientists. The diversity of the animals and the depiction of their behaviors make for fantastic fiction, with all the ingredients of a novel. The suspenseful scenario has a love story, children who are by turns enthralled and terrified, good and evil characters losing their lives in horrible fashion, and strange beasts that can be both engaging and scary. Like Saint George slaying the dragon, man always ends up vanquishing the dinosaurs, even the frightening *Velociraptor* and *Tyrannosaurus rex*. The dinosaurs are excellent stereotypes of movie villains. Huge and terrifying, they frighten children. But children need not fear them, because dinosaurs no longer exist. *But they did exist!* It is this fact that distinguishes this tale from other Hollywood creations in which extraterrestrials, with totally imaginary forms, battle man.

Thanks to these films, a whole generation has become interested in the life and death of these huge reptiles that today are but fossils, fascinating organisms of another age that left so many traces on Earth before disappearing suddenly. These films have therefore become good tools to sensitize the public to part of the history of life.

Regarding the dinosaurs, the question all moviegoers ask upon leaving a darkened auditorium is what caused them to go extinct. An abundant scientific literature accompanied the appearance of these films. We learned from it that paleontologists have proposed several theories for the sudden dinosaur extinction.

Among the most banal of these is a particularly virulent bacterial or viral epidemic. The most surprising is that of a sudden exposure of the dinosaurs to an episode of global warming that overheated their testicles, sterilizing them.[7]

The hypothesis of a giant meteorite falling to Earth first appeared in 1980.[8] A rocky asteroid, or comet of ice, exploded upon striking the Earth, raising a cloud of dust and cinders that enveloped the whole planet, causing a long, dark, glacial winter. Most vegetation disappeared as a consequence of this sudden climate change. This could explain the death first of vegetarian dinosaurs, then of carnivorous ones that fed on them.

This hypothesis was buttressed by elements of a meticulous scientific inquest. Closely examining rocks from Italy of the epoch in which the dinosaurs disappeared, a geologist was intrigued by the existence of a fine layer of clay embedded between calcareous layers derived from plankton. The lower layer marked the end of the Cretaceous, and the upper layer marked the early Tertiary, characterized by a severe loss of marine biodiversity. Aided by physicists, he found in the clay an element that is very rare on Earth but common in meteorites: iridium, a metal closely related chemically to platinum. Afterwards, this element was discovered in a hundred different sites around the planet in the clay trapped between the two sedimentary layers. Each time, the same anomaly dated to 65 million years, the epoch during which all dinosaurs disappeared. This major finding led to the idea of an explosion of a meteorite upon impact, an explosion that would have left a signature all over the planet: iridium dust.

This hypothesis was later supported by the discovery of a crater left by a meteorite impact: a gigantic cavity today partly submerged in the Gulf of Mexico near the Yucatan Peninsula (fig. 22a). The impact that formed this crater also dates to 65 million years ago.

Scientific calculations allow estimation of the force of this impact. The extraterrestrial meteorite more than 10 kilometers in diameter (a mass greater than that of Mount Everest!) must have hit the Earth at a speed near 100,000 kilometers per hour. The impact would have caused an explosion equivalent to that of 100 billion tons of dynamite, a force ten thousand times greater than that of the entire nuclear arsenal of the planet when nuclear stocks were greatest.

In the impact zone, everything was pulverized. It has been established that there were two fire blasts: the first was extremely hot, forming a cloud of vaporized rocks; the second, not as hot, was generated by limestone ($CaCO_3$) that burst into flame, liberating carbon dioxide (CO_2). A vast area was devastated by a gigantic tidal wave. The violence of the impact

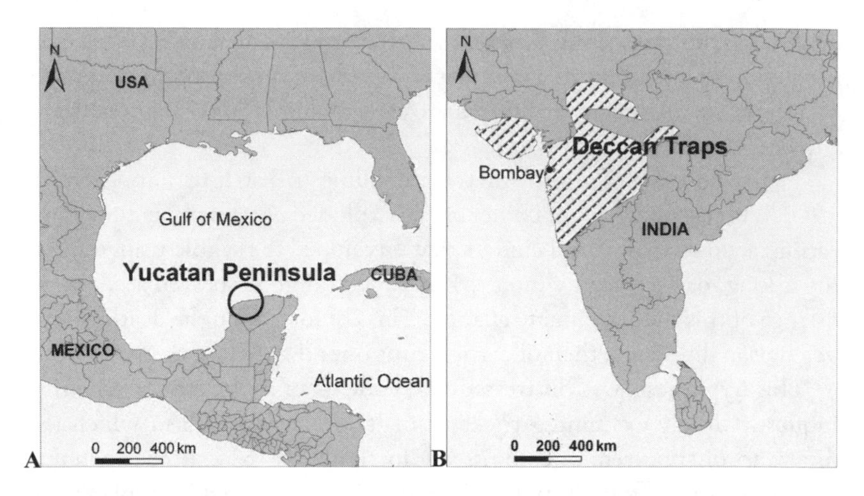

Figure 22. *A*, Map of the Yucatan Peninsula with the zone of the impact of a large meteorite dated from 65 million years ago. *B*, Region of the Deccan Traps in India, where enormous volcanic activity occurred 65 million years ago. (Drawings by Alexandre Meinesz, arranged by Fabrice Jarel.)

made the entire Earth tremble and raised a thick black dust that darkened the atmosphere for a long time. Everything that lived on our planet was disrupted, if not destroyed. Many species were thus crushed, incinerated, drowned, frozen, or starved.

This was the result of the 1755 Lisbon earthquake on a global scale: a fourth of the species that existed on Earth before the impact disappeared forever. Among the species or whole lineages that were extinguished were the dinosaurs, these enormous monsters that could not find shelter or establish reserves to combat the unforeseeable, deteriorating conditions.

Who has not imagined a similar situation today? Of course, our astrologers would see the arrival of the asteroid as depicted in an episode of the Belgian cartoon *The Adventures of Tintin* entitled, "The Shooting Star."[9] We could try to explode it in flight (a large scale "Star Wars" attempt), deflect its trajectory by exploding a well-aimed nuclear bomb . . . but if the mass of the asteroid is too great, we would have to do our best to protect ourselves and accumulate reserves to subsist for a long time. Obviously, only a small fraction of humankind would have the means to survive. Mines, active or not, would be sought out as refuges, but they would be monopolized by the dominant figures of our societies and the military. I once had a dream about digging out a shelter under my house and thought about all I would store in it to save my family: air regeneration system, food, water, books, an energy source for light and heat, vitamins and other medications. Then I imagined myself imitating Noah, taking the most minimal sample possible of life, useful or not—insects, seeds, small birds, rodents. The dream then became so complicated that I awoke, surprised to have elaborated such a strategy.

The hypothesis of a meteorite impact is today the best known, because it is the best publicized by the popular media. However, there exists an alternative theory just as plausible and supported by many pieces of evidence; Vincent Courtillot, professor of geophysics in Paris, advances it vigorously.[10] According to him and some of his colleagues, the dinosaurs disappeared following titanic volcanic eruptions. Bubbles of incandescent material from the depths of the earth burst to the surface, creating gigantic outpourings of lava (more than 2 million cubic kilometers) and gas (billions of cubic meters of carbon dioxide, gaseous sulfuric acid, hydrochloric acid, and many other toxic compounds). This state persisted for several tens of thousands of centuries (a wink of the eye on a geological time scale) in western India (in the region of the Deccan plateau), creating mountains of basalt (fig. 22b). Nowadays we can still make out the successive layers of these terrestrial outpourings; they look like giant stairways. These formations are as large as France or Texas, and they date from the epoch of the dinosaur disappearance.[11]

The two hypotheses are well supported and explain cataclysmic events that occurred in the same epoch. We must acknowledge that both events happened and they surely greatly increased the number of species extinguished on Earth when the dinosaurs disappeared.

Planetary Catastrophes of the Past: Revolutions for Evolution

In fact, the cataclysms of a meteorite strike and volcanic activity that destroyed the dinosaurs should be considered to be two cases of violent mass destruction of the fauna and flora among several past planetary catastrophes. The French naturalist Georges Cuvier (1769–1832), a founder of the comparative study of vertebrate fossils, was one of the first to depict these mass extinctions eloquently and passionately. He elaborated a theory of "revolutions" (this was the era of the French Revolution and the word was popular) that explained species extinction.[12] According to Cuvier, at several times, colossal catastrophes (revolutions) eradicated most life on this planet. He was persuaded that after each catastrophe a new creation occurred, improving on the previous species. For him, each time, the newly created life arrived at a new equilibrium. Cuvier did not believe that species could evolve from one into another. In his view, after each cataclysm, the survivors and the newly created species proliferated without ever changing form: creations added to life on Earth, and all species were frozen or fixed through time. This visionary of the mass extinction of species never accepted the theory of transformation advanced by his contemporary, Lamarck.

For Cuvier and many of his colleagues, who were subsequently called catastrophists, it was obvious: everything was buried by floods several times. Each revolution was caused by a deluge just like the one the Bible describes, followed by a drastic drying of the coasts. As evidence, Cuvier pointed to the surprising location of many fossil marine organisms in high mountain areas, suddenly inundated by waters very long ago. For Cuvier, the biblical flood was the last of these revolutions (still unknown then were mechanisms of movements of the Earth's crust that shifted and uplifted the seafloor, with all the fossils contained in its sediments, to produce mountains or make continents drift).

Though we now understand much better the characteristics of the development of microbes and the unicellular and multicellular fauna and flora since their first appearance on Earth, we also know their evolution has indeed often been disrupted by another kind of luck: cataclysms.

From 3.85 to 3.5 billion years ago (the apparent epoch of the appearance of life on Earth) until the sudden emergence of many traces of multicellular species beginning 570 million years ago, violent cataclysms were frequent. During this interval, the atmosphere was thin and allowed asteroids to penetrate more easily. These masses that remained unattached after the formation of the sun and the planets took some time to amass in a stable group of similar solar orbits where they currently move in great number. But during the first billion years after the formation of the Earth, they were still fairly numerous everywhere in the solar system, and some of them collided with the Earth and other planets. Today we can clearly see scars of impacts of this astral bombardment on the moon and Mars, where surface crusts have long been fixed.

At the same time, the amplitude of glacial cycles—temperature fluctuations—went well beyond those associated with warming and cooling episodes that have affected life on Earth during the last 600 million years. Some authors describe a "snowball Earth" in which the oceans were frozen from the poles to what are now the temperate zones, and sea levels were so low that the continental shelves, the cradle of life, were nearly dry.[13]

Moreover, the Earth's crust, composed of moving plates of juxtaposed rocks, was still very unstable. The heart of our planet, undergoing fusion, burst to the surface in many places. Devastating earthquakes and gigantic volcanic eruptions accompanied continental drift, here and there compressing the crust to create mountain ranges.

I explained in chapter 6 that, from the appearance of life on Earth (before 3.5 billion years ago) until around 570 million years ago, only nomadic, microscopic life could have proliferated in the oceans. During each catastrophic episode a minority of these planktonic living forms must have survived. During this very long period, how many promising bacterial species were crushed by cataclysms? How many combinations of bacteria, forming early animal and plant cells, were eradicated despite their avant-garde constitutions? How many newly sketched multicellular forms, roots of potentially dominant lineages, were cut down? No one can answer these questions. We will never be able to reconstitute the future that a destroyed life-form might have produced in the course of history.

Between Two Cataclysms, Visible Life Unfolds in the Oceans

Between 670 and 570 million years ago, the Earth was transformed several times into a ball of ice. Each time, life had to retreat to narrow oases of temperate waters situated near the equator. It was at the end of this series of severe climatic crises that the first animals left easily seen traces. Conical sponges are the first animals identified; they were attached to the seafloor 580 million years ago. But the first animal fossils are especially well represented by a coherent ensemble of living structures, strange and visible to the naked eye. These were mostly shaped like large biscuits (some are one meter long), rounded and regularly striped (fig. 23a–d). Their traces—simple imprints—were first discovered in the Ediacara region of Australia;[14] paleontologists subsequently observed that sediments from other regions of the world, made of fine beach sand, later aggregated into sandstone, mostly between 570 and 540 million years old, contained similar traces. Many of these strange forms of life resemble flat corals.[15] Some of them were possibly some type of jellyfish crawling over the narrow continental shelves. These supposed large flat corals or jellyfish then disappeared abruptly 540 million years ago from most of the geological archives (only few of these flat, large fossils date from 510 million years ago). What happened to them? Lethal glaciation? A meteorite? Volcanic activity? Whatever the cause, most of this entire ensemble of unusual, emergent life was eradicated.

Figure 23. Fossils of the first large multicellular animals. *A–D*, Ediacaran fauna from 600 million years ago. *A–C*, different forms from South Australia (*A*, *Dickensonia*; *B*, *Tribrachidium*; *C*, *Mawsonites*). *D*, Large decimetric imprints of Ediacaran organisms (*Pteridinium*, ca. 560 million years old, from Namibia). *E*, Examples of small shelled fossils. These are the first fossils (dating from 540 million years ago) of calcareous skeletons or coverings of animals (e.g., mollusks and sponges). (Photos *A–C*, courtesy of Pamela J. W. Gore; photo *D*, courtesy of Shin-ichi Kawakami; photo *E*, courtesy of Li Guo Xian Li and Jean Vannier.)

In rocks that date from several hundred thousand years after the disappearance of the dominant Ediacaran fauna, we find another group of organisms whose fossilized calcareous remains resemble small shells, tubes, or ice cream cones (fig. 23e).[16] This fauna, first found in the vicinity of the Siberian town of Tommot, is also widely distributed over the planet. These living structures also left vestiges, then disappeared without leaving similar descendants. Why?

The abundance of fossils found in other rocks dated from 540 to 515 million years ago seems to indicate that life suddenly "exploded" in

Figure 23. (*continued*)

diversity. Two major independent sites with exceptional fossils were discovered nearly a century ago. In 1910, French geologists mapping the Chinese province of Yunnan (then under French administration) found strange fossils at Chengjiang but paid little attention to them; 75 years later, Chinese paleontologists began to describe these fossils. More than 150 species of very different animals have been found at Chengjiang. This fauna, which flourished between 540 and 520 million years ago, shows that littoral biotic communities composed of animals and multicellular algae were already highly structured then.[17] The animals were filter-feeders (sponges and corals), decomposers (worms), and predators (crustaceans); among the latter were forms similar to vertebrates.

Another site bears witness to this sudden proliferation of multicellular life some 15 million years later (dated from 515 million years ago). It is in the vicinity of Burgess in British Columbia, Canada. This deposit of extraordinary fossils was also discovered in the early twentieth century. It is located on the flanks of a bare mountain that rises to 3,000 meters and was formerly the base of a lagoon surrounded by steep slopes. The Burgess schists, thinly sectioned, yield natural treasures. Many bizarrely shaped animals, differing greatly from each other, look like they are imprinted in the rock.[18] The first paleontologist to discover them, Charles Doolittle Walcott (1850–1927), gave them new names and classified them among the known groups of species. He called the thin cylindrical ones worms, classified the ones that resembled little shrimp as crustaceans, and the bowl-shaped ones as jellyfish. The Burgess fossils represent a surprising degree of biodiversity, for all the major body plans of today's fauna coexist there. The history of nature was suddenly enriched with faunal remains as diverse as the sponges, corals, worms, crustaceans, echinoderms (the lineage of sea urchins and starfish), and an ancestor of vertebrates (represented at that time by a little worm with a dorsal nerve cord).[19] The fossils, once they were described and classified, were deposited in museum drawers. Sixty years later, it took great perspicacity on the part of other researchers to question the classification of several of these species.[20] Even if several erroneous interpretations remain in the remarkable work of redescribing these fossils, we can clearly distinguish among the Burgess bestiary some fossils that correspond to groups of animals whose body plans are totally unknown nowadays.[21] If they were found alive today in some isolated cave, we could very well believe that Martians have landed. However, the Burgess animals lived 515 million years ago; they ate algae or hunted.

There remain only fossils, humble traces on gray stones, a cemetery of lineages and intermediate evolutionary forms that have disappeared forever. I saw some of these fossils in the Smithsonian National Museum of Natural History in Washington, D.C. I was surprised by their size; those I could see were but blackened imprints of minuscule animals representing species very different from those we see today. Some are chimeras—monsters that are half shrimp, half worm; they have five ocular globes and a long trunk that ends in pincers (fig. 24b).[22] From these little beasts with bizarre shapes and mysterious habits, novelists and filmmakers would be able to imagine remarkably original tales. But they are both too supernatural and too small, and they are therefore insignificant in the eyes of the general public; they are bad actors for such scenarios. We have to concede that, among the ranks of extinct animals, the dinosaurs, enormous, strange, expressive beasts, are ideal protagonists to arouse our interest and emotions. Their history will always be the best known.

It was again Stephen J. Gould who awakened the public to the knowledge of the bizarre Burgess fauna. He was excited about this page in the history of life and described it in detail in one of his books.[23] Correctly, he repeatedly recalled that this past diversity is the most extraordinary proof of the *contingency* of life. Contingency! This is the word that applies precisely to the whole of the evolutionary process. The Burgess bestiary wonderfully illustrates the blind luck and the cobbling together of the living edifice that rises on the enormous cemetery of aborted attempts. Were it not for, perhaps, the impact of a nearby meteorite, depending on its precise position and that of the earth at the time, or for the fatal competition with another species, some little creature of the Burgess

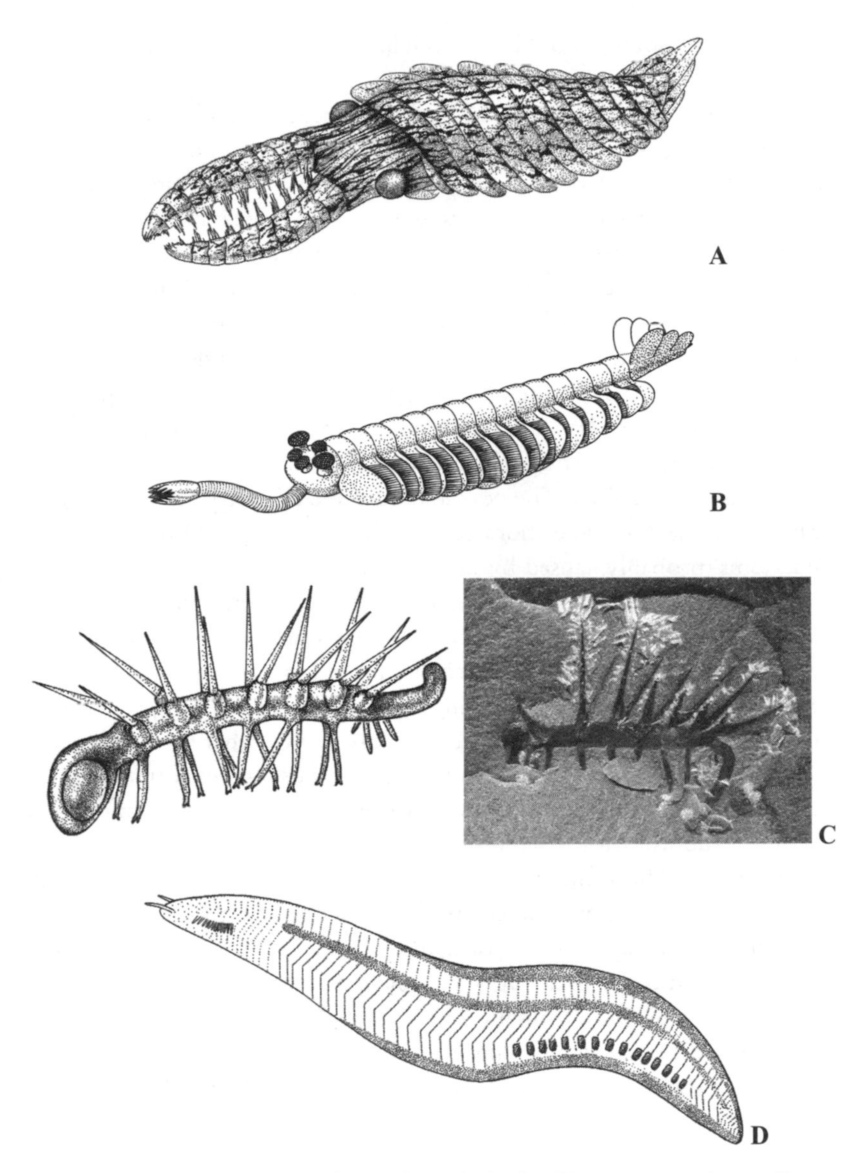

Figure 24. Seemingly supernatural forms of animals that lived between 520 and 505 million years ago, the fossils of which were found near Burgess, Canada, and Chengjiang, China. (Drawings courtesy of Mary Parrish, with permission of the Smithsonian Institution.) *A*, *Anomalocaris* (which probably grew to a meter long). *B*, *Opabinia* (6.5 centimeters long). *C*, *Hallucigenia* (drawing and blackened imprint in the Burgess rocks; 2.5 millimeters long). *D*, *Pikaia* (the earliest chordate, 4 centimeters long).

deposits—and not us—could very well have become the dominant animal on Earth.

After the great surprise caused by the descriptions and analyses of the fossils of Chengjiang and Burgess, researchers noted that many forms were found in both sites. Although the continents have been greatly displaced in the intervening 520 million years, we must recognize that a very diversified biota, similar from site to site, developed over large regions during that period. Descendants of the diverse Chengjiang and Burgess faunas have been found fossilized in younger rocks scattered around the world. Subsequently, dominant lineages emerged. They include an increasing number of species, and some of them gave rise to the great evolutionary branches that exist today. This great random base of the evolution of animals of visible size arose during this epoch. But the march of evolution has been occasionally halted. The new structure and diversity of life achieved, or in the course of being elaborated, was destroyed several times. The five cataclysms (probably caused by immense volcanic eruptions, meteorites, or climatic changes) of 435 million, 355 million, 252 million, 215 million, and 65 million years ago have left the deepest scars (these dates have been continually refined by improved geochemical and geophysical techniques). Some of these crises have destroyed 85 percent of all marine species. Each time, the surviving species diversified, evolved into other forms, through selection for new adaptations to available habitats, and eventually a new cataclysm occurred. The dominant species can disappear, leaving room to former subordinates that, in turn, undergo the same fate because cataclysms recur. This is how the dinosaurs appeared following a cataclysm caused by the impact of a meteorite that struck the Earth a bit more than 200 million years ago. Traces of iridium have also been found in rocks dated to this time of great upheaval of biodiversity.[24] Then, about 65 million years ago, the dominant dinosaurs yielded their place to rodents, which until that time were an insignificant and dominated animal. These, by continuous evolution, produced a genus of apes, then man.

Planetary Cataclysms: The Fourth Bit of Luck

We must recognize the importance of these cataclysmic events and assess their crucial impact on the history of life. We must understand that the three layers of contingencies—mutations, genetic mixing, and natural selection generated by environmental circumstances—do not suffice to explain life today and its history.[25]

The blind luck of when and where cataclysms occur is independent of contingent genetic changes that are the basis of evolutionary innovation (mutations and genetic mixing). Even though genes have a single objective, that is, to reproduce themselves in living descendants, and employ all their stratagems to persist, they are at the mercy of a flick of fate. No matter what its genes, an organism will be destroyed if it happens to be in the unforeseeable path of a comet or asteroid or near a massive volcanic eruption.

The genes that exist today are not those that have succeeded after several billion years of evolution only because they were "better"—they succeeded above all because they were lucky enough to escape the cataclysms.

Unforeseeable cataclysms have effects different from those caused by the random chance of the environmental situation that modeled the species. They are not part of the fortuitous circumstantial causes leading to the natural selection of species adapted to environmental constraints. In effect, only the Rube Goldberg evolutionary constructions that are neutral or advantageous for the multiple but relatively stable "normal" environmental constraints will last and proliferate. The organisms that are endlessly selected by the features of the environment in which they evolve are confronted with a fourth layer of random events. At any moment, they can be faced with total destruction, unpredictable and definitive, however well adapted they are to the usual fluctuations of their environment.

It is thus that three types of randomness that have continuously molded life by successive improvements of adaptations to environmental constraints (mutations, genetic mixing, and natural selection) can be considered as positive, because they are creative and innovative. The result of their actions is submitted to the fourth form of randomness, which is negative (fig. 25). These are the blind, unpredictable catastrophes that indiscriminately strike actively proliferating lineages that were otherwise destined for a good future and lineages that were declining. It does not act by refining but by massive destruction. It is an *antigenesis*.

This duality was incorporated very early in religions and philosophies. The spiritual interpretation of the genesis of life has always been associated with divine creation, while the destructive cataclysms have been considered as the malicious actions of the devil or punitive actions of God, such as the biblical flood.

In our era of genetics and molecular biology, we are excited about the chemical mechanisms of life and surprised to discover the many contributions that reductionist science at the molecular level can bring to an understanding of the history of life. Nevertheless, it is important not to

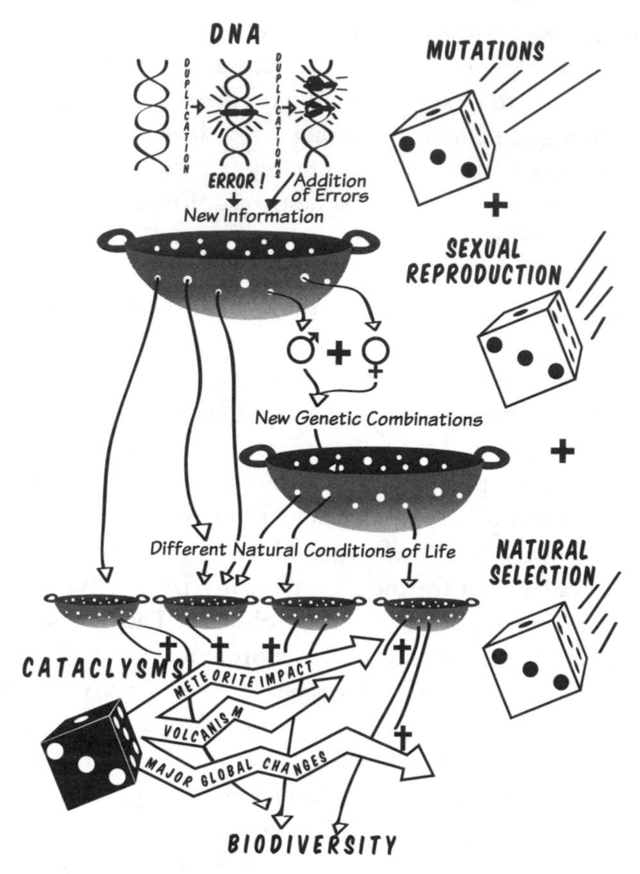

Figure 25. Evolution: the fourth form of randomness. The three positive, creative, and innovative random events of mutations, genetic mixing, and natural selection, and the negative random event of major cataclysms. The colanders represent the physical and chemical constraints that filter out particular innovations from the mass that arise. (Drawings by Alexandre Meinesz, arranged by Marjorie Meinesz.)

ignore essential elements in the pathway of life that do not fit this model. Effects of mutations, genetic mixing, and natural selection on evolution have been and remain subjects requiring profound theoretical and technical approaches. Effects of cataclysms on evolution are like death in a car accident: easy to understand, requiring little scientific work to comprehend the ends of some lineages. Just because it is easy to understand car accidents does not mean we should neglect their occurrence or their effects on life expectancy.

The blind luck of cataclysms—"accidents"—has played a determining role in the history of life; it should be placed on the same pedestal as the three other indeterminate successions of contingencies in the hierarchy of mechanisms that have selected the living beings found on Earth today.

The remains of the biscuit-like *Ediacara* animals and those of dinosaurs, not to mention the vestiges of the "quasi-extraterrestrials" of Burgess, remind us that the elaboration of biodiversity on Earth was unpredictable, without a predetermined plan. Nothing is set, nothing can be fixed in advance. Life is dynamic, it continues to change with many disruptions: the severe periods of extinction followed by dazzling regeneration.

The Biblical Flood, A Local Cataclysm

Each time, nature has gotten over it.

When all of this first seemed evident and clear to me, I thought of the Bible, and of what the minister had taught my daughter. In the first texts of the Old Testament is the story of the first men who went bad.[26] They committed many sins, so God decided to eliminate these monsters, and to this end He organized a cataclysm: a gigantic flood. He saved only one old wise man, Noah, whom He allowed to be accompanied by his wife, his sons, and his daughters-in-law. In this episode of extreme eugenics, a sort of ultra-severe artificial selection, we should recall that Noah loaded his ark with many animals and plants, by pairs. And when the cataclysmic inundation finally ended, he released this little community into nature and life began anew.

How did the prophets have this vision of such an extraordinary story? It is a metaphor that superbly symbolizes reality. In fact, there were cataclysms, there were survivors each time, and each time these survivors dispersed, reproduced, and diversified.

But just as for the other parables in the book of Genesis, the story of Noah now has a scientific version. According to many scientists, an immense flood really did occur between 7,150 and 5,300 years ago in the Middle East on the banks of the Black Sea or on the shores of the Sea of Marmara. It was not generated by a staggering deluge of torrential rain coming from who knows where, but by a completely explicable, yet brutal and stupendous phenomenon.

The Black Sea is located to the northeast of the Mediterranean Sea and connected to the latter by the Dardanelles, the Sea of Marmara, and the Bosporus (fig. 26). Its waters today wash the shores of Turkey, Georgia, the Russian Federation, the Autonomous Republic of Crimea, the Ukraine, Romania, and Bulgaria. Some authors[27] argue that 7,150 years ago, the

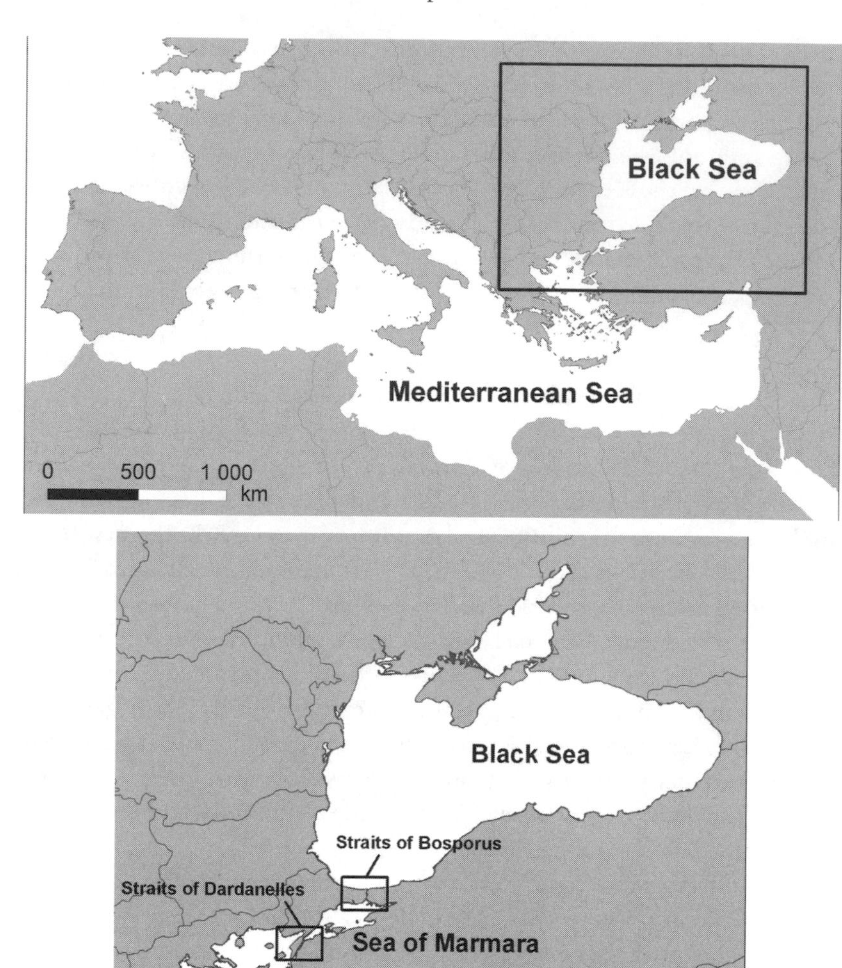

Figure 26. Map of the Black Sea and the surrounding area. (Drawings by Alexandre Meinesz, arranged by Julien Gratiot.)

Black Sea was an immense lake separated from the Mediterranean, and its surface was 150 meters lower than that of the Mediterranean (fig. 27a). For some reason (an earthquake along the junction of the continental plates of Europe and Asia and exceptional rain, perhaps?), the land separating the Black Sea from the Sea of Marmara was breached. The waters of the Mediterranean rushed into the Black Sea. The authors who gathered the evidence that proved the occurrence of this event describe an immense, roiling river following the Dardanelles and the Sea of Marmara, running

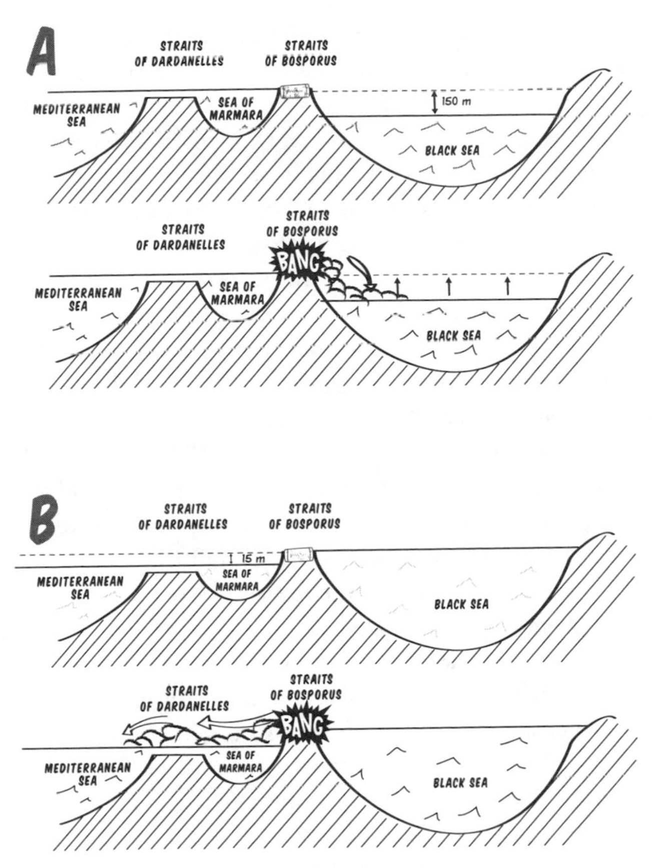

Figure 27. The scientific version of the story of Noah: an immense flood really did occur between 7,150 and 5,300 years ago in the Middle East on the banks of the Black Sea or on the shores of the Sea of Marmara. *A*, One hypothesis is that of a cataclysmic flood that poured water from the Mediterranean into the Black Sea between 7,150 and 5,300 years ago. *B*, Another is that a cataclysmic flood caused water to pour rapidly from the Black Sea into the Mediterranean, flooding the shores of the Sea of Marmara, also between 7,150 and 5,300 years ago. (Drawings by Alexandre Meinesz, arranged by Marjorie Meinesz.)

right over Bosporus and falling into the Black Sea in the form of a giant waterfall equivalent in volume to 400 Niagara Falls. Simple models of the flow of fluids between basins of different sizes show that the rise in the level of the Black Sea was relatively slow (one estimate is that it lasted for 10 years).[28]

Other authors[29] argue that the Black Sea gradually filled up with the retreat of the last glaciers (12,000 years ago), with water flowing in from the great rivers of eastern Europe, especially from the four D's: the Danube, Dniester, Dnieper, and Don rivers (fig. 27b). The level of the Black Sea gradually rose to that of the Sea of Marmara. But an immense plug of sediment from a river emptying to the west of the Bosporus separated the fresh water of the Black Sea from the saltwater of the Sea of Marmara (connected with the Mediterranean by the Dardanelles for more than 12,000 years). At some date not yet agreed upon (between 7,150 and 5,300 years ago), the sedimentary plug broke open. Water from the Black Sea then flooded the shores of the Sea of Marmara before flowing into the Mediterranean via the Dardanelles. Ever since, water flows from the Black Sea towards the Mediterranean through the Sea of Marmara.

If this hypothesis is correct, the flooding of the shores of the Sea of Marmara was sudden, because this sea is narrow and communicates with the Mediterranean only by the long, sinuous strait of the Dardanelles. It is remarkable that this hypothesis (that the Black Sea was higher than the Mediterranean, and the former poured into the latter through Bosporus) was considered by Greek authors at the time of Aristotle and Plato.[30]

Even though today we see disagreement among interpretations of many studies of the sediments, fossils, and geology adduced in support of these hypotheses, one thing is certain: an exceptional event occurred between 7,150 and 5,300 years ago associated with the communication of the Black Sea with the Mediterranean via the Sea of Marmara. It is a simple story of bodies of water at different levels that ended up communicating with one another with a cataclysmic flood of the lower water body.

At the time of this event, the region of the Black Sea and the Sea of Marmara was the cradle of Western civilization. All indications are that a developed society occupied the welcoming shores of the Black Sea and the Sea of Marmara. Faced with the sudden rise in the level of the sea flooding all the coastal cities, everyone had to flee, abandoning everything. Those who occupied the shores of monotonous alluvial plains were surely the most surprised faced with the swiftly rising waters, and they must have been the most devastated.

It is also very probable that a certain Noah survived better than the others, fleeing by boat rather than on foot and taking on board men, women, and children of his family and all the seeds and livestock that would again be useful when everything stabilized.

This history must have profoundly marked the survivors, who emigrated to surrounding lands. The real history was transformed into a legend perpetuated from generation to generation throughout the Middle East and used by the Apostles to illustrate the supernatural, divine powers that weigh on man—and especially sinners.[31]

The theme of an apocalypse wrought by God on Noah's descendants remains popular among gurus who predict an imminent end to our civilization. We can remember the fantastic predictions of Nostradamus and the ill-fated prognostications of sect leaders who, as the third millennium approached, induced their followers to commit suicide rather than face the suffering of the approaching apocalypse. All these predictions are based on the fact that a devastating cataclysm really was likely to occur eventually, and they managed to make people believe that their visionary abilities allowed them to predict the event precisely. Many were fooled. So too were the opportunistic prognosticators who subsequently exploited the terror and destitution of survivors of natural catastrophes, interpreting them as predestined to strike.

Reading *Candide* clarifies matters: apocalypses are neither programmed nor predestined but arbitrary; they are random and contingent. Moreover, Voltaire was surprised by the biblical interpretation of the flood; as a provocation, he affirmed that this was for him the greatest miracle described by the Bible.

Here is what Voltaire deduced on the subject in 1734:

> [B]ut that water should have covered the entire globe at one time is a chimera, demonstrated to be impossible by the laws of gravitation and of fluids, and by the insufficient quantity of water. We do not claim to be doing the slightest damage to the great truth of the universal deluge, reported in the Pentateuch: on the contrary, that is a miracle; hence we must believe it; it is a miracle, for it could not have been accomplished by the laws of physics.
>
> In the history of the deluge everything is miraculous: it is miraculous that forty days of rain should inundate the four quarters of the earth and that the water should rise fifteen cubits above all the highest mountains; miraculous that there should be cataracts, gates, openings in the heavens; miraculous that all the animals from every part of the world should betake themselves to the ark; miraculous that Noah should find enough to feed them for ten months; miraculous that the ark should be able to accommodate all the animals with their provisions; miraculous that most of them should not have died; miraculous that they should have found something to eat when they left the ark.[32]

To illustrate the perception of the parable of the flood, Stephen J. Gould tells the story of the Reverend Thomas Burnet (1635–1715), who was unable to believe this legend, despite his faith and his ecclesiastical position.[33] Burnet was impressed by the statement of a contemporary, "I can as soon believe that a man could be drowned in his own spittle as that the world should be deluged by the water in it." Burnet had to retract this statement; he had too accurately emphasized the improbability of the planetary flood and had thus contradicted the Bible. Gould also relates the story of the successor to Isaac Newton as the chair of professor of mathematics at Cambridge University, William Whiston (1667–1752). He developed a surprising theory according to which a comet brushed the Earth, and its tail, consisting of water, inundated the Earth, thereby provoking the flood.[34]

Whatever its basis, the reference to the cataclysm, just after the genesis, is troubling. The prophets knew absolutely nothing, nothing of evolution or of the dramatic episodes that devastated the lives that preceded those of men. However, they chose this particular story, which reminds us that life is fragile and at the mercy of terrifying cataclysms that can destroy almost everything. Even if this story for them symbolized the divine preeminence that governs our fate and that of all life around us; even if the prophets have transformed a real event into a moralistic warning to humankind not to stray from the path of righteousness—"this is what happens to men who transgress against divine laws!"—it remains true that everything can be destroyed overnight. And everything will begin anew nevertheless.

Whatever the causes, it is a fact that life was brutally destroyed between 7,150 and 5,300 years ago along the shores of the Black Sea or the Sea of Marmara. And it was destroyed on a global scale 65 million years ago. It has been devastated several times since it emerged 3.85 billion years ago. Each time, there have been survivors who drastically changed the subsequent course of evolution. The disappearance of the faunas of Ediacara, of Tommot, of Burgess, and the famous dinosaurs reinforce the concept that life emerges by a succession of random events.

Evolution: Order of Events Determined by Chance, the Opposite of a Predestined Process

At this stage of the description of major events that have sculpted life on earth, I can summarize as follows:

The interweaving of the three structuring geneses (the emergence of life—elsewhere?—with the appearance of bacteria on Earth, the origin of plant and animal cells, and the emergence of multicellular species with differentiated cells), confronted by three levels of constructive randomness constituting the evolutionary mechanisms and a layer of destructive randomness (the cataclysms), can explain the history of the diversity of life today (fig. 25).

I realize that removing the sacred myth of divine creation and a predetermined genesis of life was shocking in Darwin's day and remains so today for fundamentalists whose culture is still based on the book of Genesis in the Bible. Some of them will come to realize that it is a beautiful fable that is no longer credible, and their disillusionment will resemble that of the child who learns there is no Santa Claus. This scientific dissection of the different kinds of geneses and the luck that has shaped the diversity of life can lead believers to question the foundations of their faith. It can also rock the thoughts of philosophers and romantics who prefer to imagine life as something other than the result of a long, pitiless, random struggle for survival in the face of incessant arbitrary decapitations.

Just as Voltaire sensed for the lives of men, we must integrate the cruel reality of a fundamental role for luck and contingencies in the emergence of life, including that of man. The fortuitous events, fundamentally unforeseeable and unexpected, are sometimes for the better, sometimes for the worse. They are not earned, whatever our attitude and our past. This is why the domain of contingency is hard to accept in the context of traditions that allow us both to justify the hope of better days and to see the hand of fate or punishment in unfortunate events.

In fact, the Bible should not be considered a manual of the history of life (though many fundamentalist creationists consider it as such).[35] Everything in it is symbolic, mythic, moralistic, spiritual, sacred. Those who keep their faith while recognizing the blazing progress of the sciences in casting into doubt one of the pillars of religion (the genesis of life) believe that ancient, divine messages can be adapted to modern knowledge. They can accept that, when the first monotheists lived in Egypt several thousand years before Jesus Christ, the divine word could not make Moses, Abraham, and their disciples understand that the embryo of life is a DNA molecule representing a code, that there exist microscopic lives invisible to the naked eye . . . which, by addition, have produced fauna and flora that evolve thanks to mutations and genetic mixing, with some organisms subsequently changing into multicellular forms shaped by physical and chemical constraints and selected by other environmental exigencies.

The scientific revelations of Van Leeuwenhoek, Lamarck, Darwin, Mendel, Crick, Watson, and so many others can be considered as so many victories in the progress of knowledge that the divine word could not reveal to the ancient authors of the Bible, Torah, Koran, or the sacred texts of the Asian faiths on the origin of life. Nowadays, many religions tend to integrate in various ways the rapid advances of scientific knowledge that must be increasingly acknowledged even by the faithful. To interpret the biblical words indicating that God created man in His image, various religions now propose that He mainly gave man intelligence.[36] They tend also towards the belief that man was essentially destined to accumulate the knowledge to do a better job of understanding, maintaining, and managing what God produced.[37] In this sense, these religions view science and knowledge as a divine evolutionary heritage that should be acknowledged and developed in order to orient our ethics and behavior towards present-day life (our demographic pressure and the erosion of biodiversity attributed to our overpopulation is called the *sixth mass extinction* by some authors[38]). This trend has been forcefully expressed by the American biologist Kenneth Miller at the dawn of the twenty-first century.[39]

This is a different way of interpreting scientific progress than that of proponents of the popular idea of intelligent design. These recognize the succession of steps widely recognized as having occurred in evolution (e.g., the passage from bacterium to man), but they deny the role of contingency or chance in this evolution. This spiritual theory is based on the belief that God directs all evolutionary processes according to a preestablished intelligent design. Intelligent design has therefore become a metaphysical or religious alternative to creationism, which has become increasingly difficult to defend. This theory was developed in the United States during the 1990s, and belief in it was recently encouraged by President George W. Bush, who has advocated its inclusion in the high school science curriculum. By contrast, belief in intelligent design appears to trouble some Catholics. In November 2005, the Rev. George Coyne, a Jesuit astronomer and director of the Vatican observatory, unambiguously rejected the theory of intelligent design. In his view, it is unscientific and should not be taught alongside scientific subjects. He added that this idea is not spiritual; he does not wish us to consider God as a "Newtonian God who made the universe as a watch that ticks along regularly." However, at roughly the same time, Pope Benedict XVI seemed to encourage the teaching of intelligent design by affirming that the universe was created as an "intelligent project" and criticizing "those who in the name of science say its creation was without direction or order."[40]

This spiritual alternative to creationism shows that scientific advances have had a major impact on religious beliefs. Religious dogmas adapt to scientific discoveries, and ancient writings are no longer read literally: we try to interpret them. Similarly, New Age gurus establish new lines of metaphysical thought camouflaged in scientific language. However, whatever the evolution of these beliefs and thoughts, the spiritual and scientific magisteria will always remain separate. For adherents of the spiritual magisterium, there will always be belief, faith. I only observe that believers can integrate scientific progress without losing their faith.

Most scientists are insensitive to the spiritual implications of their discoveries. They will continue to accumulate new evolutionary knowledge, testing hypothesis after hypothesis, rejecting them or tentatively accepting them. But at a time when the history of the evolution of species is better and better known and is becoming universally recognized, it is on the matter of the determinism of the process that the arguments have become the most heated. Some scientists ascribe ever-increasing importance to the physical and chemical constraints that govern evolution. The colanders in figure 25 represent the constraints that filter evolution. For these theorists,[41] physical and chemical constraints inevitably lead evolution to produce convergences of form or function. This reasoning leads to the suggestion that the pathway that led to the evolution of man was inevitable. Although these scientists are careful to designate theories that are more spiritual than scientific as intelligent design, they are proposing a view of evolution in which the role of chance is reduced, where necessity reigns, canalized by various constraints that reduce the scope of evolutionary possibilities.

With respect to this debate, it is good to recall the theme of this chapter: the action of global cataclysms on the tranquil pathways of evolution. It is the simplest example that best illustrates the fact that evolution is fundamentally contingent; this is the most comprehensible explanation for the public at large. It is the arbitrary choice imposed by sudden cataclysms that so troubled Voltaire two centuries ago, for there is really no logic to these major events; they have arbitrarily destroyed lives with promising futures. These cataclysms have given contingent orientations to the evolution of life in a brutal, illogical manner, without reason or intelligence. To consider cataclysmic contingencies as one of the four layers of chance in the elaboration of life on Earth (along with the contingencies of mutation, genetic recombination during sexual reproduction, and natural selection) should reinforce the view that the design of life follows no logical course. Nothing can be predetermined or fated in the face of such events as have often ravaged the environment and the diversity of life on our planet.

Similarly, we have shown that it has above all been unions that have enlarged the scope of life. Some of these unions occurred in the distant past in very special circumstances. These events have not recurred—there is no longer spontaneous generation of bacteria from molecules, there are no more original unions of bacteria creating new combinations of life. These unions were opportunistic, contingent associations. If they had not happened, or if other types of unions had happened, biodiversity would be very different today.

Whether the source of an idea is spiritual or scientific, it is important to note that the fundamentally contingent and deterministic aspects of evolutionary history have now become the source of the most contentious confrontations of ideas between the two magisteria. In any event, faced with the sequence of troubling, inexplicable coincidences of the pathways of life that seem to be strung together by fate, nothing should keep us from remaining optimistic and enthusiastic. The true stoic accepts fate without considering it as programmed or earned but retains free will with respect to all that he or she can affect.

Look again at *The Astronomer,* that superb composition by Vermeer. Nothing was left to chance by this perfectionist. Every detail suggests something. Behind the model, Antoni van Leeuwenhoek, spirituality is represented by the subject of a painting on a wall. In the shadows, we can barely identify this work. But art historians have noted that the same painting, in large format, figures also in another Vermeer work, *Lady Writing a Letter with her Maid*. Behind the lady writing a letter and her standing servant, the same painting, well lit, occupies the entire wall. It represents Moses saved from the waters.

By extraordinary circumstances and luck, the infant Moses survived the madness of both man and the waters of the Nile that carried the fragile basket in which he had been placed. Of course, sacred books explain that, as for Noah, it was God who governed Moses's destiny. This is an allegory of divine providence, that of God the Protector. Moses later shows the right path to his monotheistic people who lived in Egypt. He is also given traits of an astronomer, insights in a science well developed in ancient Egypt. The prophets and Vermeer knew how to reconcile the spiritual and scientific magisteria.

CHAPTER 9

The End of the Evolutionary String

The Tempo of Evolution, and Now . . .

Representing Evolution in the Distant Past

I often reflect on pedagogical approaches that can make students understand the immeasurable time during which life developed on Earth. How can I render the notion of distant time more concrete, and how can I establish in their minds the moments of the geneses and great advances in the evolution of life? Most authors use a clock for this purpose. They reduce the length of time from the origin of life on Earth until today to 12 or 24 hours and explain that, on this time scale, man appeared only in the last two minutes.[1] This temporal image of evolution annoys me, because it gives the impression that time is contained in a cycle. It continues to flow, but it remains stuck in a 12-hour clock, so it seems like an elastic band that can contract. This is why I prefer the image of a string to that of a clock—a string whose length represents the time that has passed. The emotional effect is the same: man does not arrive on the scene until very close to the end of the string.[2] But to my eyes, the string better represents time, which flows in a linear, unidirectional fashion, like an arrow. This is sometimes called *sagittal time* (derived from the Latin *sagitta,* which means arrow)[3].

I must explain that, in this image of time, there is no beginning. Past time is infinite. The human mind has difficulty conceiving of this notion of ancient eternity. To suggest this feature, one end of the string should be hidden in a ball of twine of a size that does not allow an estimate of the length. The string of time has only one end, that of the present time, the one that has an arrow indicating that time continues to flow. To represent the location in linear time of the geneses of life on Earth, it is necessary to unroll the ball enough to free up a suitable length of string. A major event should serve as a landmark to give a scale to the string. For that we can choose the big bang, which is theorized to have occurred 15 billion years ago. The visible, unrolled string therefore represents 15 billion years. But with this choice, it is too difficult to mark on this string the great steps in the development of life, because they would be too concentrated in the last quarter. To limit the string of past time to a more recent period, we can choose as a starting point the time when our planet was formed. This was about 4.56 billion years ago. This gives the string a good scale for the temporal representation of the evolution of life on our planet.

But even with this arbitrary limit to the length of the distant past, I know that these few billions of years constitute a period so vast that it is an abstract notion.

In fact, the time that we have a good sense of is that which is determined by the position of the sun and the alternation of day and night. It is the time given by our watches. It is the time indicated by the Gregorian calendar (instituted by Pope Gregory XIII in 1582), a system that divides time into parts: days, weeks, months, and years. It is based on the solar cycle and is marked by the passing of seasons, of equinoxes, and of solstices (the Egyptians adopted a version of this kind of time before 4000 BC). Similarly, time is understood concretely by reference to the cycles of the phases of the moon on which the Jewish and Muslim calendars are based (the lunar crescent is the symbol of Islam). The Babylonians were the first to strive to construct a lunar calendar.

At the scale of one human generation, time is made concrete by anniversaries of the births and deaths of persons close to us. To make time in the more distant past concrete, we use historical markers related to the (approximate) date of the birth of Jesus Christ and the cycles of the sun. In France, the dates of the last world wars (1939–1945 and 1914–1918, respectively), the French Revolution (1789), and the Battle of Marignan (1515) are the best known. But for earlier events, the dates blur in our minds. In earlier chapters, I gave numbers of years like 10,000, or 100,000, a million, or a billion; even those figures, for most of us, do not signify a concrete length of time. Only geologists have managed to integrate into their consciousnesses the very distant past, by breaking it down into eras (primary, secondary, and so forth), periods (such as the Cambrian, Ordovician, and Silurian), epochs (like lower and upper Cambrian), and other subdivisions distinguished by the nature of their rocks, sediments, or fossils. Without knowing all the subtleties of the many geological ages, we can use a simple string to visualize the temporal position of ancient times and to have an idea of how much time separates them.

Before testing the string metaphor with my university students (an audience already well acquainted with the subject of evolution, but uncritical), I tried it out in a lecture to an audience poorly acquainted with the subject but who knew me well and therefore felt free to criticize me: my Lions Club.

My subject was the dates of the geneses. There were about fifty persons at the dinner lecture, members with spouses and friends. They were all waiting to see me hold forth on complex topics projected by modern means: video or overhead projector. But I had brought only a paper bag decorated with drawings of carnival masks and streamers (it was Mardi Gras). From the bag, I took out my string. I enjoyed disconcerting my audience, surprising them. The effect was immediate. There were several friendly gibes on the theme of the precarious state of university professors who were reduced to presenting a subject with a piece of string. I then stretched the string nine meters between the two opposite walls of the room. The restaurateur, worried by my actions, betrayed his nervousness at this unusual opening of a lecture in his establishment. My friends again expressed their great amusement; some of them tried to reassure our host.

When I finished stretching the string, I changed my tone.

"This is deep time—time that is hard to grasp."

I then removed from my bag two sheets of paper and two clothespins. One the first, which I attached by one of the clothespins to one end of the string, was written in large letters "Today." At the other end of the string, I attached the sheet labeled "Formation of the Earth: 4.56 billion years ago" (fig. 28).

Then I asked the following question:

"How many years does one meter of string represent?"

The calculation is simple: if nine meters stands for 4.56 billion years, a meter of string represents about 500 million years.

I therefore hung on the string little colored pieces of paper on which were inscribed the round numbers: 500 million years, then 1, 2, 3, and 4 billion years.

Then I positioned chronological markers for the geneses of the great steps of evolution, beginning with the most recent and most concrete and ending with the oldest and most abstract.

"Let's first locate the Cro-Magnons, the people who lived in Henri's cave [the Chauvet Cave; see chapter 1]. As you know, they were there about 30,000 years ago."

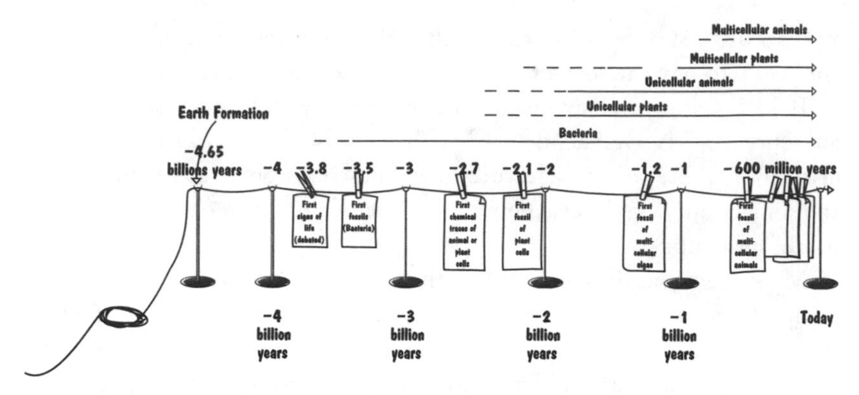

Figure 28. The "evolutionary string," which helps people grasp the enormous span of time between the first appearance of life on Earth (bacteria, between 3.8 and 3.5 billion years ago) and the emergence of multicellular forms visible to the naked eye (600 million years ago). (Drawings by Alexandre Meinesz, arranged by Marjorie Meinesz.)

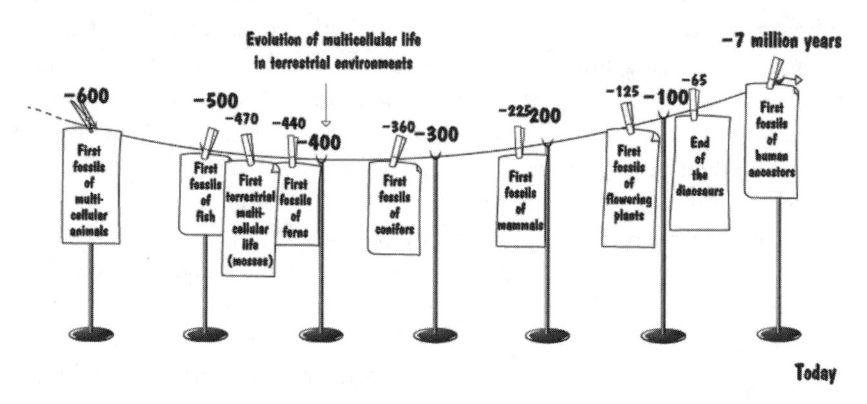

Figure 29. Detail of the "evolutionary string." The best-studied part of evolution occurred over the last 600 million years, with the beginning of the colonization of newly emerged dry land by multicellular organisms visible to the naked eye (mosses and crabs) and including the appearance of the first prehumans 7 million years ago. (Drawing by Alexandre Meinesz, arranged by Marjorie Meinesz.)

I then went over to the end of the string marked "Today."

"You see, one meter equals 500 million years, so one centimeter equals 5 million years, and one millimeter corresponds to 500,000 years. So it's impossible to attach the paper marked 'Cro-Magnon' correctly!"

I therefore attached it with adhesive tape to the "Today" label.

Then, I took out of my bag the label "Appearance of the first prehumans, about 6 or 7 million years ago." I hung it with a clothespin just one centimeter from the "Today" label. (fig. 29).

"Let's make a big leap into the past, to an event we all know: the disappearance of the dinosaurs 65 million years ago."

I hung this label 13 centimeters from the "Today" label.

Then I continued by taking out labels for other important events and positioning them in the right places on the string.

At 26 centimeters came the label "Appearance of flowering plants, 125 million years ago."

At 45 centimeters: "Appearance of first mammals, 225 million years ago."

At 72 centimeters: "'Appearance of conifers, 360 million years ago."

At 82 centimeters: "Appearance of first ferns, 410 million years ago."

At 88 centimeters: "Appearance of first terrestrial plants (mosses), around 470 million years ago."

At the meter mark: "Appearance of first fish, 500 million years ago."

At 1.2 meters: "Appearance of first large animals (Ediacara), 600 million years ago."

At 2.4 meters: "Appearance of first multicellular large plants (algae), 1.2 billion years ago."

At 5.25 meters: "Appearance of first microscopic animal and plant cells, 2.7 billion years ago."

At 7 meters: "Appearance of first fossil bacteria, 3.5 billion years ago."

And finally, the first chemical traces attesting to the appearance of bacteria, 3.85 billion years ago, were represented by a label 7.7 meters from the end of the string that marked today.

The string took on the appearance of a garland. I had spent a quarter of an hour decorating it in total silence. A quarter of an hour to go back through the course of deep time. The temporal landmarks were positioned at the scale of the length of string.

"Now look at the first label I've stuck up there, for the appearance of prehumans, 1 centimeter from the end. That was 6 or 7 million years ago. Isn't that an insignificant length of time in the history of life?"

I then invited my listeners to look closely at the string and to meditate on what it revealed about the time it took for life to develop on Earth. The time it took for a bacterium to transform itself by additions and evolution into a man.

"Look at the immensity of the time that passed, the time it took for life visible to the eye to emerge!"

Several of them remarked that the closer I was to the end of the string labeled "Today," the more labels were hung up. They questioned me: what happened that caused evolution to be accelerated beginning 550 million years ago?

"That's only an illusion. I have positioned in that region a series of events that we relate to. If there aren't many labels on the largest part of the string between the appearance of bacteria and the appearance of many animals and plants visible to the naked eye, it is because the evolutionary events have appeared, to our eyes, rare, minor, or insignificant. In fact, life never stopped evolving, whether the evolution occurred inside organisms composed of a single cell or among the vast assemblage of cells that make up macroscopic species. This last process that I mentioned, spectacular in our view, didn't happen until the last quarter of the time life has existed on Earth."

Then I went back over the times of the geneses to explain the most remarkable evolutionary events, first taking out of my bag of tricks a set of Russian matryoshka nesting dolls to explain the successive additions leading from bacteria to the origin of animal and plant cells (see chapter 4); then a cylindrical thermometer containing a liquid in which temperature is indicated by spheres of differing buoyancy, some of which rested on the bottom while others floated on top or stayed at intermediate levels (the ones that indicated temperature), to illustrate the severe constraints imposed on the initial cells living in water (see chapter 6); some Lego pieces to explain the formation of multicellular organisms (see chapter 7); and finally a coral of the genus *Fungia* that resembles the first fossil animals visible to the naked eye (from the era of fossils of Ediacara). I exploited all the contrasts, between the latest knowledge and the simplest means of communication, between such serious matters and my bag of tricks decorated with the carnival figures. I had been inspired by Vermeer's painting, depicting Van Leeuwenhoek admiring a miniature image of the universe: a celestial globe. He had used the same method as the string to perceive infinity: the heavens and the galaxies were reduced to a sphere that he cradled in his right hand. Thanks to this reduction of scale, Van Leeuwenhoek, who discovered microbes, was able to dream and reflect on the order of the world, from microscopic life to the infinitely large heavens. But he always kept his hand firmly resting on the table . . . on the materialistic basis of everything here below.

The Three Geneses of Life

The image of the string is good at allowing an understanding of the amplitude of the time needed for the development of life on Earth. It permits us to integrate better the dynamic of life and to imagine the situations of the distant past: everything was not always as we see it today. Nothing has been stuck in one place since the beginning of time. Life was not created in a day or in seven days; it has been transformed, it has evolved for a very long time.

With this book, I have tried to distinguish three fundamental stages of life, beginning with its origin on Earth (fig. 30). These three successive, distinct geneses were:

- the appearance of bacteria.
- the formation of microscopic animal and plant cells.
- the elaboration of multicellular forms composed of different cell types.

The emergence of self-consciousness in our species might be considered a fourth genesis, more spiritual than biological (this is the *noogenesis* of Teilhard de Chardin).

Each of the three geneses left Earth with a type of life that subsequently evolved on its own. Thus, bacteria have produced many other types of bacteria, the independent, nomadic plant and animal cells have evolved internally and produced hundreds of thousands of different unicellular species. Finally, the multicellular forms have produced the diverse flora and fauna that is visible to the naked eye. Each of these stages has been countered in its drive to diversify by a number of blind, unexpected, destructive cataclysms.

It has therefore been little by little that life came to flourish in profusion. Sedentary multicellular animals acquired effective means of locomotion and spread in the vast ocean pastures composed of unicellular organisms. About 450 million years ago, life visible to the naked eye left the water to conquer land.

Among multicellular organisms, a collection of large species, multiple assemblages of cells in harmony with the ambient physical and chemical forces, tested by natural selection and memorized in the form of genetic information, allowed the step-by-step emergence of different architectural and functional equilibria.[4] Some algae became mosses, some of these became ferns, some ferns evolved into seed plants and then flowering plants. Some crustacean lineages became insects, some fish left the water to become amphibians or reptiles, some reptiles became dinosaurs, then birds, others were modified into mammals, some mammals became apes, and certain apes became humans.

The exciting aspects of the causes of evolution of forms within a lineage (usually respecting a basic functional architecture, like many car models all based on the presence of four wheels, a motor, and a passenger compartment) and changes in the functioning of an organ or cell have been the subject of an abundant scientific literature on evolution since the revelation of the transformation of species by Lamarck at the beginning of the nineteenth century.

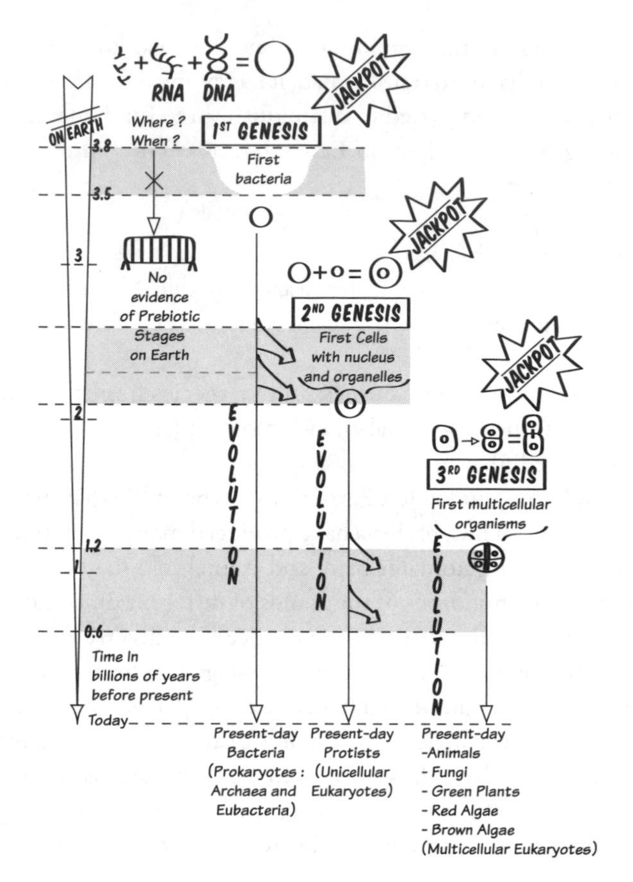

Figure 30. The three geneses of life. Each genesis derives from a union that produces new living organizations (evolutionary "jackpots"). The fruits of the three geneses coexist today on Earth (bacteria, protists, and the lineages of multicellular eukaryotes), while prebiotic assemblages have never been found (there is no documented spontaneous generation of bacteria). (Drawing by Alexandre Meinesz, arranged by Marjorie Meinesz.)

Hierarchy of Evolutionary Mechanisms

The links are studied, they are explained by increasingly well understood mechanisms, but evolutionists still debate the relative importance of the major causes of great evolutionary changes.

To clarify these debates, the German paleontologist Adolf Seilacher defined three groups of factors that could guide the construction of organisms (the elaboration of forms, or *morphogenesis*).[5] By analogy, Stephen J. Gould extended these ideas to functions and behaviors, and therefore to the totality of evolving life.[6] In a simplified summary, these authors propose the following clusters of constraints:

- historical constraints (the store of genetic information that changes by mutation and gene flow and that gives specific traits to descendents).
- structural or architectural constraints (the sum of physical and chemical forces tending to orient forms towards logical architectures independently of any specific functional adaptation).[7]
- functional constraints that adapt an organism to its environment (resulting from natural selection favoring those who survive and reproduce better than others in a particular environment).

These three groups of constraints of development and evolution are diagrammed as an equilateral triangle, with each vertex representing a group. All causes of the dynamic and direction of evolution are therefore considered equally.

Arguments between biologists center on the dominance of one vertex of the triangle over the others. Responding to the dogmatism of partisans of dominance by the functional vertex (adaptation governed by natural selection), Stephen J. Gould, with his colleague, Richard C. Lewontin,[8] observed an error in certain authors' reasoning. They criticized those who separate a living being into characteristics (or traits), then propose for each trait taken alone a sequence of evolutionary events leading to that trait that seems logical and "for the best" in the sense of optimizing adaptation to the environment. For such authors, each facet of an organism is the result of a *program of adaptation,* the steps of which were selected by nature to lead to the "best" state: towards the perfection of the whole organism. Gould and Lewontin criticize partisans of such an evolution, driven essentially by natural selection on individual traits leading inevitably towards improvements that, when cumulated, lead to forms, functions, and behaviors that are ever more sophisticated.

The Metaphor of the Central Cupola of the Basilica of San Marco in Venice

To show the fallacy in such an interpretation of evolution, Gould and Lewontin used a surprising and tortuous metaphor: the central cupola of the basilica of San Marco in Venice (which has since become famous among evolutionary biologists worldwide). They first observed that the decorators of the basilica had used four triangular spaces surrounding the cupola to depict four key personages. These spaces, available to the decorators, are pendentives, designated by Gould and Lewontin as spandrels: that is, in a larger architectural context, spaces above an opening (door or window) or between two arches.[9]

The spandrels Gould and Lewontin used as examples were elaborated by eleventh-century builders in neutral fashion—they were unavoidable architectural elements associated with the construction of the cupola and the four pillars that support it. The decorators, coming along after the builders, used the cupola and the four spandrels to create a magnificent composition of Byzantine mosaics conceived entirely to respect a four-way symmetry. This is why the cupola and spandrels seem, a posteriori, well adapted to their decoration. In fact, the presence and form of the spandrels as well as the artistic composition of the mosaics are epiphenomena, secondary productions, byproducts, works of secondary consequence. First of all, the architectural constraint (to rest a cupola on four load-bearing columns) imposed the presence of spandrels; next, the existence of these spandrels suggested a decoration adapted to their shape. There had never been a plan by the architects and decorators to create spandrels and to decorate them. The magnificent decoration of the gilt mosaics of San Marco were never the raison d'être of the basilica.

Gould and Lewontin's message was twofold. First, it is important not to use inverse reasoning—to reverse cause and effect. Although one can easily admit that the decorators did a great job of adapting their mosaic images to the form, number, and arrangement of the spandrels, it is important not to claim that the architecture was adapted to the spandrels or to the decoration. In order to clinch this fact, Gould and Lewontin also referred to Voltaire, who used *Candide* to demonstrate the absurdity of the optimism of Pangloss, who was an advocate of the theory that everything that exists was foreseen, and everything therefore works toward a final goal. Gould, Lewontin, and other authors cited other, simpler examples to illustrate inverse reasoning. Thus, the nose did not evolve to support eyeglasses, oranges were not adapted so that humans could see them easily and collect them from trees, the stripes on a melon are not adapted to facilitate its division into slices for a family. These very examples were advanced by eighteenth-century authors in a positive sense, because they were persuaded that the Creator had always arranged matters for the best. In *Candide,* Voltaire caricatured them in the person of Pangloss, who endlessly proclaimed that all was for the best in this best of all possible worlds.

Secondly, with the example of the spandrels, Gould and Lewontin showed that structures can arise as secondary consequences of (and not as direct adaptations to) another decision (in the San Marco example, the decision was the construction of four columns to support a cupola).[10] Living organisms can evolve exactly in this way: structures, colors, functions, and new behaviors can arise as indirect consequences of natural selection for other characteristics. This demonstration emphasizes the importance of two of the three vertices of Seilacher's triangle (historical and architectural constraints). It shows that the series of successive and cumulative adaptations is not the sole mechanism of evolution. Adaptation does not always have at its origin a gradual series of unidirectional changes selected for by the environment. In other words, adaptation should not be considered as consisting exclusively of a constant refinement caused by natural selection always pushing all the different parts of an organism to be better and better, with the final objective of a perfect organism.

This double metaphor—Pangloss and the spandrels of San Marco—has caused much ink to flow in the small world of evolutionists. Controversies over architecture (the domes or cupolas, the spandrels or pendentives) and over the notions of adaptation and selection have been the subject of several books and many articles.[11]

Therefore, on a subsequent trip to Venice, I was intensely curious about these famous spandrels. I noticed that the edifice is composed essentially of five main cupolas, each with four spandrels. Each group of spandrels was shrewdly decorated with mosaics. Each architectural space (cupolas, spandrels, arches, vaults, lunettes, niches) was used "for the best" by the decorators to represent a biblical scene with mosaics. Gould and Lewontin cited and depicted only one of the cupolas with its four spandrels: the one in the center (called the cupola of the Ascension), the one in which the iconography is best divided by quadrilateral symmetry. Raising my eyes to admire these places of great religious significance (and, more recently, of great significance for evolutionary biologists), I remembered the first sentence of Gould and Lewontin's publication. According to them, "The great central dome of St. Mark's Cathedral in Venice presents a detailed iconography expressing the mainstays of Christian faith." That does not do the basilica justice. Among the five main cupolas, that of the Ascension is the one best chosen to make reference to Christianity. All the main personages of Christianity are shown: Christ, Mary, the twelve disciples, and the four evangelists (one per spandrel). Under the central cupola, while admiring the marvelous mosaics in this environment that marked the history of the Mediterranean region for seven centuries, and under which Galileo came to meditate, I smiled, also remembering the title of Gould and Lewontin's publication: "The spandrels of San Marco and the Panglossian paradigm: a critique of the adaptationist programme."

This title suddenly appeared to me somewhat provocative, if not surreal. Gould and Lewontin chose the most symbolic and venerated images in the sanctuary of Christianity in which all the figures of the New Testament are represented. To this sacred iconography, they added Pangloss of Voltaire, an author famous for his anticlerical opinions . . . and all this to explain a scientific version of evolution!

Even though Gould subsequently wrote that the choice of metaphors was apt (for the intended demonstration) and not consciously ironic (with respect to related deductions about the metaphors), I nevertheless think that they must have been having fun by deliberately choosing this symbolic example and an abstruse, convoluted title for their publication.[12]

Instead of the spandrels of the basilica of San Marco, Gould and Lewontin could have chosen a more accessible example (for didactic purposes), one more neutral and better known to researchers opposed to their theory, especially in the United States and Great Britain. For example, they could have used the monumental statue of the great presidents of the United States on Mount Rushmore in South Dakota, a national monument in their own country, to assail the errant reasoning of some of their colleagues. In fact, it was nature (tectonic uplift, erosion) that conferred a particular form (cliffs) on this granite mountain. Men considered this configuration and the characteristics of the rock in selecting this site on which to carve the traits of the American presidents. The tourist who sees the monument today could wrongly believe this mountain was preordained—adapted—to receive these busts.

The strong point of the metaphor of the spandrels of San Marco is that it demonstrates that, if something appears well adapted to the way it was used, it may nevertheless be an epiphenomenon rather than the result of a successive series of gradual adaptations. A neologism has recently been proposed to designate secondary adaptations that result in this way: *exaptations*.[13] A biological example is often cited to illustrate this concept: penguin wings. These were originally the same organs that are shared with all birds and that evolved by a long series of improvements—adaptations for flight. But for penguins, the uses of wings have changed; they aid in swimming. This is a secondary adaptation that could not have been predicted when wings were originally evolving—an exaptation.

Today another more didactic metaphor can be advanced to use the terms *adaptation* and *exaptation* correctly; this is based on the evolution of cell phones. Initially, they were large and heavy and lacked screens. By a long series of technical adaptations, they have become small and lightweight with increasingly large and luminous screens. Images on the screens were initially in black and white, later in color. These last few years, cell phones have co-opted a new function: digital photography. These two functions (telephone and digital camera) nowadays seem indissoluble and have even become interactive (photographs are sent by telephone). So well integrated to the technical architecture of the telephone, the digital photograph is a derived function, a by-product, an "epi-usage" with respect to the cell phone. The two dynamics of evolution are well illustrated by this metaphor: a program of successive adaptations led to "the best" (the miniaturization of cell phones), and an exaptation, a co-opting of a new function (digital photography), became associated with the phone (phones now incorporate other exaptations—video recorders, MP3 players, and the like).

The example of the spandrels of San Marco, associated with Voltaire's character Pangloss, is typical of some sophisticated scientific authors who use complicated metaphors not easily understood by the lay public. Gould and Lewontin used velvet gloves to assail an error in reasoning by using esoteric metaphors that only a few experts could understand. In this case, a major shift in context was needed concerning Byzantine art, Voltaire's prose, and, for non-Christians—a majority in the world—the New Testament (to understand the full subtlety of the demonstration, one would have to know there were four evangelists: one per spandrel).

In fact, this all signifies once again that the major transformations in the history of life do not respond to any one law; they are consequences of equilibria resulting from interactions of the three vertices of developmental constraints described first by Seilacher, then Gould. If we bear in mind the absence of a single universal law of evolution and take into account the several mechanisms that can explain natural complexity, evolutionary changes in fauna and flora can be explained case by case.

Transformation of Species Through Time, and the Lifespan of a Species

On the scale of a human lifetime, we also perceive periodic successions of cyclical biological events. The inexorable disappearances of individuals—ourselves, those close to us, pets, house plants—represent events in normal life cycles that we often regret. We also observe this type of event at the level of species.

The entity called a specific "species" corresponds to an ensemble of individuals who resemble one another and reproduce in natural conditions, leaving fertile offspring. The horse and the donkey are species. But the mule, resulting from hybridization between a horse and an donkey, is sterile—it is not considered a species. All humans are individuals of the same species; all horses are also part of the same species. As individuals are ephemeral, so are species:

- Some disappear without leaving descendants.[14]
- Others are gradually transformed into different species.[15]
- Finally, others are transformed into diverse varieties.

After some time, these varieties no longer have anything to do with the ancestral population; they do not interact or reproduce with one another. The varieties end by becoming intersterile, thereby becoming distinct species.[16]

To illustrate this last evolutionary tendency, I can cite the classic example of dogs. In fact, the dog species is in the process of differentiation into multiple varieties by virtue of artificial selection practiced in many places by human beings for several millennia. Thus the Chihuahua is a miniature dog and the German shepherd is a large one. Theoretically, they are interfertile, but they have diverged so much that it is now virtually impossible for them to copulate and therefore to reproduce in natural fashion. This is how several species have disappeared after having transformed into a swarm of varieties that became new species (like the several finch species Darwin studied in the Galapagos islands).

Thus species always end by disappearing, even if only by being replaced by one or several descendent species. Such a succession of species through time constitutes a lineage of species, each derived from the preceding one.

In such successions, what is the life expectancy for a species within the lineage? Archives of fossils show there is great variation. The most exhaustive study gives four million years as the average lifetime for fossil species[17]. However, although the majority of species do not undergo striking morphological change for 1 to 10 million years, others have persisted without much visible change for nearly 100 million years.[18] Several filamentous bacteria species (probably cyanobacteria) that construct stromatolites hold the record: they seem to have remained unchanged for 3.5 billion years.[19]

Another question comes to mind: in the heart of a lineage, is the persistence of species through the geological ages variable? The logic of evolution leads us to think that the accumulation of beneficial changes would lead to ever more perfect species, increasingly well adapted to their environment. More perfect forms should survive longer.

This is not the case. Analysis of many ancient lineages with fossil representatives allows us to reconstitute their evolution through the geological ages. The lifetimes of adjacent species in a lineage, successive species through time, do not increase as evolution proceeds.[20] In fact, every newly arising life form must integrate itself into the surrounding sea of life. If this entourage evolves, then this form must also evolve or change. All the component species of a community are therefore entrained, with various rhythms, in an endless race. They must continue to evolve without respite, always adapting to new situations. Evolution is not a race for progress with the goal of achieving perfect form and function. Rather, it is a race for survival. At all levels, a life-form must find its way in the context of competing life forms and changing environmental conditions. This process leads to diversification of evolutionary options. This is the reason many species change rapidly. In this context, being large and complex does not insure survival in perpetuity—just the contrary! The group with the highest fraction of species endangered, en route to extinction, is the one at the top of food chain: large predators. Large, complex, specialized in their means of subsistence, they are threatened by the simple reduction in numbers of organisms in lower levels of the food web and by human destruction of their habitat.

Size and complexity are not the ultimate goals toward which the evolution of all lineages tends. Many bacteria and unicellular animals and plants have evolved greatly all the while remaining microscopic; these are the dominant organisms of our planet. Even for multicellular species, to become more complex is not a fixed rule that guarantees success. Parasites often evolve by reversing or greatly simplifying the traits evolved by their distant ancestors. The important thing is effective originality—deviation from the normal in a way that improves survival or reproduction.

The egoism of the competing species constituting life on Earth entrains an endless evolution. It is always relative, opportunistic, fertile, and innovative.

The Rhythm of Evolution

The rhythm of transformations has also been the subject of many theories and heated arguments. In fact, aside from dramatic periods in which large fractions of the ensemble of living organisms are all threatened by cataclysms, species have evolved at diverse rates.

Darwin believed in slow, gradual change—a species produces one or more other species, which change gradually through time before disappearing, perhaps after surviving for awhile in a parallel fashion with each other.

But in layers of superposed rocks, we often find anomalies. Frequently we observe great thicknesses of rocks containing fossils—petrified vestiges—that resemble one another and in which forms change almost imperceptibly through time. Then, suddenly, at the level of the next higher (more recent) layer, fossils of a new species appear, related to an older one but clearly distinct from it. Such observations led an early-twentieth-century scientist, Richard Goldschmidt, to the concept of *hopeful monsters* created by huge, sudden mutations (causing an individual to "jump" from one species to another) well adapted to a changing environment.[21] Increasingly rare defenders of this theory of evolution, which has today fallen into disfavor, have been termed saltationists.

Studying other superposed layers of marine sediments rich in fossils that petrified over the course of thousands of millennia, other authors (including Niles Eldredge and Stephen J. Gould) were struck especially by the very long periods when no evolution of fossil forms was evident (this fact had been explained by paleontologists as resulting from the absence of fossilization of intermediate forms between notably different species in a lineage). These epochs, which lasted several million years during which evolution was more or less stalled, are called periods of *stasis*.[22] Evolution has been blocked in a nearly perfect state of equilibrium for a long time. This equilibrium is *punctuated*—interrupted—by the extinction of species and the sudden appearance of new forms (often clearly related to their predecessors) that end by dominating the community and overthrowing the equilibrium. The main contribution of Eldredge and Gould underlined the discontinuities in the dynamic of evolution.[23]

This concept of a dynamic evolution can lead us to think that a sudden acceleration of the pace of evolution is sometimes unleashed for no apparent reason. The newly evolved forms upsetting the equilibrium should not be interpreted as being the fruits of reproductive anomalies or mutations (producing in one fell swoop macroevolution or hopeful monsters in the midst of a population). They arise from small populations at the margins of a species' range in which a high proportion of individuals are different genetically from individuals in the main populations. Isolated on the margins, they can maintain these differences. They reproduce and end up replacing all the populations in a relatively short time. These expatriates, because they are initially embedded in a small population, are able to increase the frequency of their particular genes more rapidly than their cousins in the large ancestral population. When they subsequently invade their region of origin, they can provoke an upheaval. If these new species are better adapted than their relatives to the environment they now all inhabit, they will become dominant. In this case, the equilibrium or stasis ends, punctuated by rapid change that establishes a new equilibrium.

In fact, all these variations in the tempo of evolution are possible and occur on a case-by-case basis. The rhythms and cycles of the evolution of life are fundamentally irregular. They make me think of the surface of the sea. It can be calm, analogous to the long stases in which changes are slight. It can be dominated by regularly spaced small waves, analogous to slow and gradual evolutionary transformation. And when a storm arises, this is the analog of the cataclysm, in which everything that has built up is suddenly destroyed. Flux and reflux of life on Earth.

In the swelling waves that constitute the cyclic succession of individuals and species, some disappear forever without leaving descendants. This is the fate for many species that have come in contact with man. The best known include the saber-toothed tiger, the mammoth, and, more recently, the famous dodo (a large, flightless bird, shaped a bit like a fat turkey, that lived peacefully on Mauritius until the first explorers eliminated it to feed themselves), and the Tasmanian wolf, among others, for the list of species extinct since the arrival of man is very long. Birth and death of an individual and of a species are inherent in the nature of life. They are necessary punctuations in the natural system, waves of life forever breaking on the shore of time.

This logic has long been unacceptable (and remains so) for man, who has sought a mythic explanation as much for the creation of life and its diversity as for death, especially our own and that of our loved ones. It is therefore remarkable to observe that the main religions all hold out the hope of eternal life or at least a recycling of our souls in other lives. This is the forlorn hope that new life cycles are associated with preceding ones on time's arrow.

Similarly, this logic has become increasingly unacceptable as mankind accelerates the rhythm of punctuations. A growing number of species are disappearing because of us. Our growing population, our activities, and the pollution engendered by them constitute the beginning of a cataclysm for the life that surrounds us: a major, global upheaval for all life.[24]

Altogether, the increasing diversification of life results from an immense suite of evolutionary events that occurred with various rhythms and cycles. All this has happened thanks to billions of combinations of biological and environmental events spread out over 3.85 billion years on Earth. Nothing was predestined, nothing could have been foreseen; life today is the fruit of chance and competition. If we could rewind the film of life to its beginning and allow the first living actors to express themselves as they wish, the new film would not be like the old one, we would not be there, other species would inhabit the Earth. In fact, after each rewinding, there would be a new scenario. Nature is eccentric, unpredictable, and versatile. It is the fruit of the trials and errors by the ever-expanding multitude of the living.[25]

Current, Perceptible Evolution

Changes can also be seen in communities of living organisms. The compositions of communities are in harmony with environmental conditions, and the environment and the community can change together. All of us have surely seen the beautiful renditions by artists who have imagined what the Earth looked like in the age of giant ferns. The many fossilized remains allow us to reconstruct these landscapes of ancient life realistically. They show that alternating waves or arctic and tropical weather have ceaselessly disrupted animals, plants, and ecosystems. Individuals (you and I), species, and ecosystems succeed one another while being transformed . . . everything changing in different rhythms.

At our time scale, that of one human generation, we can see a series of natural ecosystem changes on a great scale. The fastest are found in the seas, on the shores. To illustrate the dynamic of life on a human time scale, I will describe a scientific expedition I participated in to the island of Tatoosh (fig. 31). This island is located near one end of the Strait of Juan de Fuca, less than 200 kilometers from Seattle or Vancouver. The island is a speck of confetti in the ocean, a rock several hundred meters across, encircled by high cliffs. At its summit scrubby vegetation thrives, hiding military buildings either abandoned and in ruins or well preserved but hermetically sealed for a long time. This was an outpost of a Maginot line during the time when Americans feared a Japanese attack. The island has another important feature: it is in the middle of an American Indian reservation. The leader of this expedition is someone I greatly admire, Professor Bob Paine (plate 7a), very well known in the world of naturalists

Figure 31. Map showing location of Tatoosh Island in Washington State. (Drawing by Alexandre Meinesz, arranged by Fabrice Javel.)

for his ecological experiments.[26] Strapping, avuncular, and charismatic, he is passionate about this island, the theater of experiments that made him famous. It is his favorite place in life, which has been devoted more to observing nature than to pursuing other personal interests. He was able to negotiate the study of this speck of an island with American Indian chiefs.

I went there with him during the summer of 2000, and I observed that the American Indians still have some control over these lands.

Distrust of Yankees is always strong in the remnants of the ancestral domain that the American Indians now inhabit. Paine is a good white man, a scientist bringing only wisdom. He has been given free rein to camp there as he pleases with his students and guests. Camp is the right word. Two ramshackle wooden barracks serve as shelter. One of them has been arranged as a common dormitory. It was gutted, opened to the wind. Paine has done nothing to ameliorate the situation. We need to deserve the nature that surrounds us, showing the American Indians that the scientists visiting the island are satisfied with the spartan conditions of the place. I took great care to place my sleeping bag so that I would not be covered with the droppings of birds that nest in profusion under the roof. These birds squawk beginning at sunrise, about 4:30 AM, when we get up.

We have to live with the tides, because it is only when the sea goes out that we can see the incredible marine biodiversity of this site. And, during our expedition, low tide was always around 5:00 AM. Paine's collaborators, young men and women, descended the cliffs with the aid of ropes. With sliding boots, we jumped over dark fissures where the cold, raging sea could easily drown us. Each dangerous site was named for a scientist who had been injured there. I did not want to leave my name!

Immense rocks were uncovered by the retreating tide, revealing multicolored forests of splendid algae. Some were more than six meters long. Families of basking seals were draped over islets, and a strong fishy odor pervaded the site. Flocks of seabirds nested on the island. They flew out to the high seas in the morning to feed and returned to the island in the evening to rest. And the ocean, more gray than blue, roiled by a series of swells, extended to the horizon.

Bob Paine and his students then explained their study methods. They laid out their analysis of this landscape. An overview shows that the entire area uncovered as the tide goes out teems with exuberant life. They taught me to decipher this mosaic of organisms carpeting the tops of rocks and the shadowed depressions, the zones that were low or high in relation to mean sea level. These coherent ensembles of species occupy as many distinct marine ecosystems. They are almost like organisms themselves, intimate associations of interdependent life. The many species constituting a community are either mutualistically associated or competing, for either space or food, in a perpetual war. I already understood this and had seen similar assemblages while diving in the Mediterranean or while standing on the Brittany coast. But at Tatoosh what surprised me were the mussels. Enormous mussels—it took two hands to hold some of them. The mussels tirelessly filter seawater containing a great richness of unicellular algae, a real "soup" that they drink and constantly filter when they are underwater. Huge starfish are common on the mussel beds, which measure tens of square meters. Only some particular species of algae manage to grow on this animal carpet. Between the zones covered with mussels we saw surfaces totally occupied by algae of all different sizes that formed distinct strata. Gigantic sea urchins browsed this multicolored vegetation, themselves becoming the preferred food of sea otters.[27]

The masses of mussels grow until they are smashed by the battering of the waves or of logs driven by them. The sea ends by tearing off clusters of mussels. This is when the forests of algae are able to grow in the cleared space. Young mussels then settle among the algae, end by dominating and replacing them, and the cycle repeats itself.

The site had the traits of a dynamic being.

I then grasped the right analogy for this ecological laboratory: it was the equivalent of a mosaic of fields, moors, scrub, and forests composed of interdependent species in harmony with the environmental conditions. At Tatoosh, what was spectacular was that in just several years, one could see the forest collapse, fields take over, scrub establish, and the forest return. There, it was no longer a question of the history of an individual or a species but that of an entire structured community of species. Each species (like the mussel) comprises a population of individuals whose number and density are in harmony with the environmental conditions and the other species in that habitat (the environment in which they can grow).

Ecosystems are legion on our planet, and they change in cyclical fashion. Even during a period of stable climate, they change endlessly in immutable cycles. Dominant species end by collapsing, leaving the field to species that were formerly subordinate and that now become dominant, and so forth. But this schema must be nuanced. All that is needed to change the equilibrium is several more or fewer individuals of a species, a new species mixing with those already there, a species disappearing because of competition, a particularly cold or warm season, or some disturbance like a storm. The change may be slow or quick, but nothing will remain the same.

Paine's experiments were aimed at understanding the laws of these equilibria of life. He undertook an entire series of manipulations destined to reveal the roles of the main living components of the ecosystems uncovered by the tide. He scraped rocks to clear them, then watched the order in which species recolonized the bare patches. He removed sea urchins or starfish in order to learn what would happen to the rest of the species without these browsers or predators. He installed exclosures, squares constructed of plastic or copper—a powerful repellant—to prevent certain predators or browsers from penetrating the protected or denuded zone. He then followed the development of demarcated squares. For thirty years, every three months, researchers come to the island to understand the mechanisms and patterns of competition and recolonization. For thirty years, such experiments have led to the island of Tatoosh, where, season after season, Bob and his colleagues have landed in the summer by means of a frail boat and in the winter by helicopter. He has therefore been able to prove the importance of certain species that he calls *keystone* species of the ecosystem. If they are removed, the entire system quickly collapses and the past richness is not reconstituted. He put his finger on the fragility of living communities. A small perturbation that seems insignificant can make the entire edifice tumble. And in an ecosystem, several stones support the vault.

The tide came in again. We had to work quickly, scraping bare square meters of rock, picking off giant mussels with a hoe, sprinkling the rock in exclosures with a strong detergent (oven cleaner) to be sure the experiment would start at zero. After these laborious chores, I had the notion to gather several of the mussels that had been removed. Not to study them, but to prepare a good plate of mussels marinière for my friends. Bob tactfully told me that this wasn't a good idea. He gave me a behavior lesson: for thirty years, no one had eaten a single mussel or other animal found on the rocks at Tatoosh. The island is not a nature reserve, and these are not protected species; it was simply a principle of conduct that he imposed. It is important not to view the study area as a pantry, and it is important that the American Indians not be mistaken about the motivations of the scientists. Paine made it a point of honor to respect this principle, which was intuitively favorable to his continued presence on the island. The mussels that were ripped off fell in the crevices and were recycled by the system. The scientists were wise! I had to be content with a barbecue of our main food source, which came in plastic bags—marshmallows! Paine understood my confusion, and one evening he took me to catch salmon—an activity well regulated and monitored by federal and state authorities and the local Indians.

We found ourselves off Tatoosh in the frigid waves of the Pacific in a tiny inflatable boat of 4.5 meters, propelled by a 9.9 horsepower motor. I was not afraid. I was not thinking about the killer whales that cruised the area looking for vulnerable young seals. However, our little boat would have been easy prey for this predatory carnivore of several tons, whose behavior towards man I knew nothing about. That evening, a student had seen one not far from where we were. I had also not thought about a motor breakdown in this immense sea. Love of nature renders one a bit fatalistic and daring: very often in situations that should have been disconcerting, I have found myself calm and serene in a way that subsequently astounded me.

The salmon bit endlessly, striking the hook without our having to do anything. That wasn't the problem. The hooks were not barbed; the fish were simply pricked. Three out of every four escaped not far from the boat. We caught so many salmon that we could choose which ones to keep: not too fat, not too small, not protected species. We released those we did not want with a little scratch and, doubtless, a great fright. Within an hour, we had captured two beautiful specimens with just the right traits.

A fallen cedar trunk was used as a barbecue. I remember its remarkable odor when I cut it up with a hatchet. The salmon were delicious in spite of the horror I experienced watching Bob's students eat them with toasted marshmallows. I wanted to eat the eggs that one of the two fish carried, mashed with a little lemon juice. That seemed as strange to them as it was for me to see them chewing the white, scorched, rubbery masses. I treated myself to salmon caviar, they treated themselves to marshmallows. That evening in my sleeping bag, I listened to the sea—the waves breaking on the rocks and cliffs, the birds, the raucous cries of the seals, and many other sounds of a nature so wild. The American Indians must have experienced all these impressions associated with their land and their culture. I understood their attachment to their ancestral place. I also understood the aura that Tatoosh projects for all North American marine biologists. It is a sanctuary for marine ecosystems, the laboratory giving rise to so many theories on the dynamic of life, the precarious existence of ecosystems, and their fragility.

With the rapid succession of equilibria that the researchers could observe in several years in one place, I had a realistic image of what must have taken place over much longer time spans on earth. The fern forests gradually replaced by conifers, themselves eventually ousted by flowering plants. Instead of the few years of flux and reflux of the ecosystems on Tatoosh, the waves of change took place over tens of millions of years. One vegetation replaced another because of the slow heating or cooling of the planet. A suite of states ensues, and they are never quite reconstituted identically: everything changes each time. The cyclical time of the renewal of ecosystems parallels sagittal time, the linear time that flows unidirectionally.

Current Unfortunate Evolution

But we all have other notions of time associated with changes in nature. For me, they are associated with the place where I spent my adolescence. I recalled the site of beautiful, functional buildings well integrated into the environment of scrub vegetation.

It was on a Saturday long ago that my friends and I met in front of the train station. My backpack was heavy, as it contained food, water, a sleeping bag, part of a scout tent, ground cloth, and stakes. My six companions, members of my French Boy Scout troop, were on time and had forgotten nothing. After half an hour on the train, we got off at Biot, one of the smallest coastal train stations on the French Riviera. In a line, we followed the road, then a trail that never ended. We finally took a shortcut, zigzagging across the scrubland, avoiding the thorniest bushes.

The clearing was there, just alongside a stream of crystalline water. We erected the tent and played ball until sundown. Then we searched for dead wood. We cut up the largest branches with a hatchet and prepared a fire far from the trees, on an ancient hearth inside a circle of stones. We ate around the fire and discussed the peculiar subjects of adolescence, then sang a few songs. Our leader was just 16 years old; the youngest scout was 13. We often camped in that spot and in similar ones in the vicinity of Nice.

Forty years later, I returned to these places laden with happy memories. I was well aware that I was going to be surprised by changes in the landscape, because the technology center of Sophia Antipolis had been constructed on our playing field. I was nevertheless perplexed to see that everything was unrecognizable. The clearing had become a research institute. I was unable to find the clear water of the stream, and roads had replaced the trails. What moved me most was to find that people are no longer allowed to light a campfire in the evening to warm the body and soul.

In that place, for nature, time has stopped. Nature will no longer evolve there. It seems utopian to imagine these buildings one day in ruins and nature reclaiming the upper hand. And in 40 years, what will have become of the beautiful wooded countryside now on the outskirts of my city? These thoughts are banal, but they help me to understand that the evolution of life can be disrupted by man.

In my profession, we often have to reduce the scales of time and space. This allows us to understand better the dynamic of the destruction of nature's bounty by man. It is a device similar to that of evolution summarized by a string or a clock or to the terrestrial and celestial globes of Van Leeuwenhoek. We play with scales; we shorten time and space. I experienced my most beautiful vision of this method in Japan.

In Hiroshima, at the site devastated by the atomic bomb, there is now a city with well-aligned streets, contrary to those of all other Japanese cities. This is the first thing that struck me when in 1990 I visited this place, sadly famous in the history of human civilization. In a few dozen years, what had been a field of death and ruins had become a dense, vibrant city, a great, bustling, southern metropolis. Life had reestablished itself: more precisely, the life of man and his constructions. It was to this site, redolent with notions of the destructive power of man, that I was invited by Hideki Ueshima to give a lecture. He is not a large man, small even for a Japanese. His almond-shaped eyes behind light glasses, his smile, his enthusiasm, and the frequent funny faces showing his surprise have always amused me. Hideki Ueshima, despite his debonair air, is a new-age samurai, a crusader respected in his country for his activities on behalf

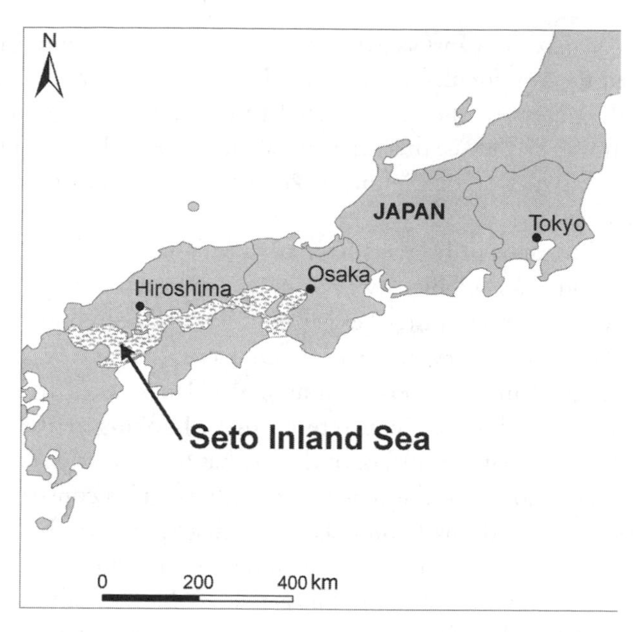

Figure 32. Map showing location of the Seto Inland Sea in Japan. (Drawing by Alexandre Meinesz, arranged by Fabrice Javel.)

of the environment. For 20 years he has directed part of a government institute that operates an immense model of the interior sea of Japan, the Seto Inland Sea (fig. 32).

The model is a vast map of water flow with detailed contours. It was constructed in an airplane hangar 300 meters long and 140 meters wide (plate 7b).[28] The currents and tides are faithfully reproduced by giant pumps that tirelessly carry up the water that flows down again to the level of two channels (Kii and Bungo) into the Pacific and one (Kanmon Strait) into the Sea of Japan. Hideki Ueshima has spent many years designing experiments to evaluate the impact of development activities planned for the 2,500-kilometer shoreline of this inland sea. How will the currents be changed? Where will pollutants accumulate? To answer these questions, he runs simulations on the model. Crossbars and cranes can liberate thousands of little floats simultaneously, to be moved by the currents. At the sites of model river mouths, dyes are injected into the water so scientists can visualize dilution of fresh water in the sea. Everything is filmed and analyzed to verify and standardize the calculations derived from the model.

I remember my first visit to the model. Hideki Ueshima asked me to follow him inside the gigantic hangar. We jumped from island to island above the sea, like Gargantuas striding over whole regions with each step. He showed me a particularly fragmented basin, Osaka Bay. He pointed out that the colors of the cement used to contain the water were not identical. It was easy to see that many parallelepipeds had been added after the initial construction of the model 20 years earlier. They represented land reclaimed from the sea.

Tens of thousands of hectares have been reclaimed as the shoreline of the Seto Inland Sea is nibbled away so cities can expand. Airports, ports, new subdivisions, and industrial expansions have all been built where sea had been. They were everywhere. From Osaka to Kobe and beyond, there is only a series of indentations at right angles, lands taken from the sea and attached to each other: concrete pustules. Looking at the model, I could easily grasp that the surface of the bay has been greatly reduced and that the great majority of the coastline is artificial. This construction has irreversibly destroyed small subtidal areas with great marine biodiversity.

The ecosystems in front of these shorelines are linear. They form narrow belts parallel to the coast. Those that are biologically the richest, which are composed of an assemblage of many species, are near the surface, where ample light penetrates. Here solar energy allows photosynthesis. These ecosystems disappear forever following construction. They are transformed into several square meters of vertical concrete walls that protect the land won from the sea. Beyond this zone, it is too deep, light is greatly attenuated, and the belt of subtidal ecosystems never reconstitutes itself in the face of the concrete.

It was in light of this observation that Hideki Ueshima explained his utter helplessness. His role was limited to advising builders so that construction modifies the currents as little as possible. But by dint of having seen bays filled in and the coast destroyed in his model, he came to understand that the environment, little by little, was degrading irreversibly by loss of surface area where most marine species could thrive. The process was never reversed; the Osaka International Airport, built in the midst of the Seto Inland Sea, would never be torn down so that nature could reclaim the subtidal spaces.

Hideki Ueshima lives in a country where many examples show the irreversibility of the degradation of nature colonized by man. On the sites of cities destroyed by giant fires (which occurred several times in Tokyo) or by devastating earthquakes (as in Kobe on Osaka Bay), man has always reclaimed his rights, never nature. For man, a disturbance is but a kick in an anthill!

Facing this vision where time and space are reduced, revealing the negative human impacts, we were silent for a long time. Then, he confided in me his disgust with the lack of foresight and overview that reigns over this matter. Each municipality, each prefecture has its pet projects that it brings to fruition without coordinating with its neighbors. Each advances its pawns on the sea in a game in which the only strategy is to foster development and win economic competitions. The builders do their best to respect the environment. To that end, they finance costly studies on the integration of projects to avoid effects of collateral pollution. But the builders seem always to ignore the essential point: the destruction of life, of communities, of ecosystems under the surface gained back from the sea. And each destructive act is added to the others, endlessly increasing the impact on coastal ecosystems. By adding up the total surface area of reclaimed land, one can easily see the damage caused to the marine environment by irreversible acts of destruction wrought over areas much greater than those that are affected *temporarily* by chemical or bacterial pollution. Despite this dismal vision that he confronts daily, Hideki Ueshima has never given up, he has never accepted that inexorable reclamation and construction will be the fate of the entire coast of the Seto Inland Sea. Patiently, and with great diplomacy, he has tried with growing success to convince the various authorities of his country that management of this body of water should truly respect the environment more. It is no longer a question of restricting his field of expertise to chemical and bacterial pollution—he advises on actions in another domain: preserving parts of the coast forever, keeping them natural. Such a strategy was developed more than a century ago in Great Britain, where private funds and bequests left to a foundation (the National Trust)[29] have allowed the purchase of coastal properties to protect them from builders' appetites. In France, a similar organization was established several decades ago: the Coastal Conservatory.[30]

Returning to Hiroshima 10 years later, I advised Hideki Ueshima to paint the borders of his model green to represent coasts that have been saved. Today, Hideki Ueshima and many other researchers aided by nongovernmental organizations urge the Japanese authorities to think more carefully about what can still be preserved.

The experimental nature of the model impressed on me the thundering dynamic of human pressure and, on our time scale, a certain irreversibility attaching to our dominance of nature.

In the Bible, it is often written clearly that our destiny is to dominate nature.

"Be fruitful and multiply, and fill the waters of the Earth, and let birds multiply on the Earth" (Genesis 1:22).

"Be fruitful and multiply, and fill the Earth. And the fear of you shall be on every beast of the Earth, on every bird of the air, on all that moves on the Earth, and on all the fish of the sea. They are given into your hand. Every moving thing that lives shall be food for you. I have given you all things, even as the green herbs" (Genesis 9:1–3).

"And as for you, be fruitful and multiply; bring forth abundantly in the Earth and multiply in it" (Genesis 9:7).

And it was done. And afterwards? What should we do now?

We have arrived at a point where nature is so thoroughly dominated that we are reducing it. The great religious founding myth of mankind's future assured by the sanctified and nourishing life around us has been surpassed. The biblical provisions stop in our time because we still multiply, but we realize that this cannot last. We overwhelm and reconstruct the Earth more and more to the detriment of vital natural resources. We dominate the fish of the sea to the extent that we have greatly reduced their stocks and threaten their very survival. Since the Neolithic, we can no longer satisfy our needs by hunting and gathering. Today, we can no longer fish for species in their natural habitats without restrictions. Such practices cannot continue.

We can no longer content ourselves with the divine oath made to Noah and his sons: God promised them that there would be no more cataclysms to destroy life on Earth.[31] The rainbow is the heavenly sign that recalls this oath to man and all living creatures on Earth. Though, in fact, no cataclysm has truly overthrown all life on Earth since Noah, life is threatened by change not foreseen in Noah's time: the effects of our own excessive reproduction!

Thus, after the questioning of the postulate that the Earth is the center of the universe, after the recognition that man is just a step in the evolution of an animal lineage, a third biblical foundation is nearing its demise. But this time it is not because of errors in the history of life as described in the book of Genesis, but because of absence of foresight.

In this context, the heralded end of the evolutionary string seems as unreal to us as its length. To forecast its future unraveling is too abstract for the mentalities of the great majority of us, and the strategies for managing our destiny are only on the scale of electoral cycles.

The evolutionary string and Ueshima's model are constructs that reduce the dimensions of space and time to render more concretely these abstract notions and to make us absorb past changes and their importance. Vermeer understood the power of the integration of dimensions and proportions on maps and celestial or terrestrial globes. The United Provinces of the Netherlands were then in the avant-garde of producing these tools. Globes and maps, so useful for the navigators of the Dutch East India Company, were produced in profusion and served to decorate bourgeois homes.

These representations are particularly well illustrated in *The Geographer,* painted as a companion piece to *The Astronomer* (plate 8a). It is the same model: Antoni van Leeuwenhoek. He has just passed a geometry examination and is posed in front of a window streaming light into the room. He seems to be reflecting on an insight that came to him as he scanned a map with the Indian Ocean. This document, fruit of so many efforts of several generations of explorer-geometricians, allows one to visualize distant lands that one discovers without having to go there. Behind him, a terrestrial globe rests on an armoire. This globe has been identified; it was constructed in Holland in 1615—that is, 18 years before Galileo was rebuked by the Pope for criticizing the religious doctrine that the sun and stars move around the Earth.

These two paintings also depict astronomy books and symbolic tools that allow the tracing of maritime routes and precise measurements of distances and positions.

In *The Astronomer,* a celestial globe and an astrolabe[32] are part of the composition. Against a wardrobe rests a poster with several clock faces apparently useful for memorizing the times of passage of stars in configurations that aid in estimating positions on the seas.

In *The Geographer,* we strain to see that a sort of cross is hanging on the backlit window. This tool—a cross-staff or Jacob's staff[33]—helped navigators of that era measure the height of stars above the horizon. But the focal point of the painting is a brass pair of dividers open at 45° that Van Leeuwenhoek holds in his hand. This tool is used to measure the distance between two points on a map and compare it to a fixed scale in order to estimate the geographic distance. Vermeer changed the painting. Originally, he had the divider points aimed at the map. Subsequently, he depicted them inclined in the same direction as the oblique beam of sunlight entering through the window. A curious point is that the lower divider point aims at a stool in the lower right hand corner of the painting. On the stool there was originally a small leaflet that he replaced by a small square.

Sunlight, the eyes of the model, the divider, and the square were all aligned. Globes, maps, the astrolabe, the cross-staff, dividers, and the square allowed scientists to have a better sense of the true situation of man in the world. They allowed better representations of the distance and time that separate the stars that were still so far away or the countries that had just been reached by the magnificent sailing ships of the House of Orange-Nassau. Now, it is this astronomer and this geographer who are so filled with wonder before the infinitesimal. We can therefore understand that a single man, Antoni van Leeuwenhoek, had embraced the quintessential knowledge of his time. The two compositions fix the context of this powerful meditation. But in these two works, Vermeer depicted the scientist with his left hand firmly posed on the table. Even while admiring the field of knowledge that he took up eagerly, Antoni van Leeuwenhoek had his feet on the ground, caught up in the material realities of the moment.

EPILOGUE

Knowing, Loving, and Protecting the Grandeur of Life

Since Cro-Magnon man, who began to draw animals in dark caves, human civilizations have admired or feared the magnificence, the power, and the complexity of life. Therefore, with respect to nature and its diversity and beauty, the sentiment that has dominated the thoughts and beliefs of humanity the longest is that the *grandeur of life* is essentially divine, supernatural.

In the seventeenth and eighteenth centuries, biologists achieved fame by astounding royal courts with the discovery of *creatures* that were exotic, beautiful, or useful for humans. Whoever brought a giraffe or rare bird from the "islands," whoever imported a tulip or a new tuber with an interesting taste was assured recognition. In this milieu, Van Leeuwenhoek's observations quickly became known; they inspired immense curiosity. The courts of Europe came to Delft to admire Van Leeuwenhoek's animalcules. Among the many visitors he received were Queen Mary II of England (plate 8b),[1] King Charles II of England, Grand Duke Cosimo II of Tuscany, and King Frederick I of Prussia. Similarly, historians describe how Czar Peter the Great moored his boat at The Hague to invite Van Leeuwenhoek aboard and how the Czar was especially intrigued seeing red blood cells circulate.

The grandeur of divine creatures seemed unlimited then; people had just discovered that it extended to the microscopic realm as well.

The Frenchman Jean-Baptiste de Lamarck was the first to dare to write that species changed through time. He was first to *admire* the phenomenon that he perceived—a gradual progression from the infinitely small to man:

> In considering first of all the simplest animal organization, which then gradually rose to that which is the most composed, as from the monad [microorganism] which, so to speak, is but an animated point, up through mammals, and among them to man, there is evidently a nuanced gradation in the composition of the organization of all the animals in the nature of the final results, which one can not only admire but that one should strive to study, to determine, and to understand well.[2]

Subsequently, it took more than two centuries for man to reconstruct a coherent history of life on earth and to change profoundly our perception of the nature of the grandeur of life.

In the conclusion of his masterwork, *On The Origin of Species,* Charles Darwin evoked the grandeur of life by referring to many natural laws that had as an admirable outcome the production of higher animals. "There is grandeur in this view of life, with its several powers, having been originally breathed by the Creator into a few forms or into one." Darwin was fascinated by the perfectionist power of the mechanism he had described; for him, "as natural selection works solely by and for the good of each being, all corporeal and mental endowments will tend to progress towards perfection."[3]

The many scientific advances since then have greatly changed our understanding of the mechanisms that have led to the splendor of the array of living species inhabiting our planet.

First, we can distinguish three fundamental steps, three successive geneses: those of bacteria, single-celled animals and plants, and finally the multicellular flora and fauna visible to the naked eye. Each step has left on our planet an ensemble of organisms that have evolved in their own ways. Bacteria appeared first, more than 3.5 billion years ago. Since then, they have continued to evolve internally, inside their single cells. Better adapted than other organisms to the different conditions of life on earth, today they dominate all ecosystems.

Next, it must be stressed that the evolutionary process of natural selection described by Darwin is just one of the mechanisms leading to the elaboration of a new species. It has been established that mutations, chromosomal mixing, and physical and chemical constraints are just as important in the emergence of a new species.

Finally, we must not forget that evolutionary forays and innovations, and living equilibria they produce, are never assured of producing descendents forever. They are always at the mercy of blind eradication caused by cataclysms.

Current knowledge also allows us to affirm that life has ceaselessly evolved without following a predestined path, without systematically orienting towards increasing complexity, without tending towards perfection. One of the great contemporary biologists of evolution, Stephen J. Gould, denounced the myth of progress in evolution and concluded that life deploys like a fan with only a passive and marginal direction towards growing complexity.[4] For him, the grandeur of life resides rather in the fundamentally fortuitous evolutionary process itself. Each mechanism that molds or selects life (mutation, chromosomal mixing, environmental conditions selecting for the best adapted, and cataclysms) is random, involving luck or contingency: "We are glorious accidents of an unpredictable process with no drive to complexity, not the expected results of evolutionary principles that yearn to produce a creature capable of understanding the mode of its own necessary construction."[5]

My analysis also sheds light on a process found in all the great evolutionary leaps: the strategy of union. At each great step of evolution, it is union that has given a new framework to life and broadened the array of living forms. The process of association and union of RNA and DNA chains is the basis for the transmissible information of life yielding the first bacteria. The process of endosymbiosis, the union of bacteria, led to lineages of animal cells, then plant cells. The process of fusional union of cells gave rise to sexual reproduction. The process of union of cells produced multicellular organisms, veritable civilizations of cells. Similarly, the union of individuals of the same species created social associations in which various social behaviors and the division of labor enhance survival and better assure the production and survival of descendents, so groups tend to be favored over individuals. Finally, the union of the first humans is at the origin of the development of self-consciousness and knowledge. It is certain that the strategy of *union* is a major part of the grandeur of life.

Thus the grandeur of life, to the extent that we can perceive it, has continually changed meaning.

Nowadays, biologists are scientific Cartesians who do not linger before the futile beauties of life. Their science, which has become very abstract, has diverged from sentiments of admiration that can be shared by all and even more from the spiritual conception of life. They can still marvel, but now these reactions are directed at a beautiful experiment, or at the logic of a mathematical model of the intimate functioning of cells. They have also been driven to dissect the mechanisms of life down to the molecular level. They have even succeeded in reconstituting the pages of the history of life by describing and comparing DNA sequences from diverse species.

Having deciphered the code that governs the architecture of animals and plants, revealed the chemical secrets of plant pigments, reconstituted the steps in the emergence and development of organisms, molecular biologists have in some sense eliminated the sacred aura of life. Everything seems mechanical or chemical to them. Disconnected from past discoveries of the magnificence of life, the revelation of the cellular, genetic, and chemical processes that elaborate life has become abstruse. These processes have therefore been somewhat devalued; the utopian goal now is to see the government enter a laboratory to explain with graphs and numbers a significant advance in one of the many specialties of biology. The value of research on life has essentially become mercantile: only discoveries that give hope of a new medical treatment or promising economic returns and patents arouse interest. Surely, to preach the grandeur of life while referring only to research on the subtle chemical mechanisms that have led to it cannot excite the masses.

This past century saw both great progress in the knowledge of life and just as great a weakening in communicating knowledge to the broad public. So many times I have wanted to show my colleagues, enclosed in their ivory towers of university laboratories, the level at which discussions about conserving biodiversity occur in the main social strata of our population. This has driven me to consider new criteria for assessing scientists, criteria that would lead them to take their science out of their laboratories. For far too many of them are too deeply buried in the depths of their knowledge and are too distant from the material realities of everyday life. This behavior is induced by competition; the only mode of professional evaluation is the number of scientific publications of high quality that will circulate only in the microcosm of their colleagues. This formal, institutional means of communication has today become morally inadequate; it should no longer be the only goal. Biologists who make discoveries and possess important knowledge should be responsible for diffusing what they have learned at a level the public can understand and should always recall the fundamentals: everything acquired in the past, the context in which new discoveries and research are interesting.

Science has progressed so rapidly that it has gotten carried away without paying attention to its roots. We should therefore not be surprised that the main religions and philosophical schools that absorb and gradually integrate scientific knowledge in their magisteria find themselves overtaken in the public eye by other messages that are easier to understand and are based only on sacred texts and secular traditions anchored in the collective subconscious. The messages of fanatic proselytes, fundamentalist or sectarian, therefore have free rein in a world in which science has nevertheless been able to answer so many questions about our past and our future.

I have been able to see the penetration of scientific progress in knowledge of life within a panel of representatives of the economic and social worlds. As I have presided for more than a decade over a commission charged with analyzing the environmental management of a region of 4.7 million inhabitants (Provence-Alps-French Riviera),[6] I have the opportunity to converse at length with many people from different social classes and with widely differing levels of education. The members of my commission agree on one point: an eagerness for knowledge on the environment and nature conservation at their level of comprehension. However, all are largely ignorant of the recent discoveries on the functioning of life and the subtle mechanisms of its evolution and decline. A single fact, recurring in the lay media, has garnered all their attention: life on Earth is now facing damage and destruction by humans, and this effect is accelerating. They are all preoccupied by increasing pollution with its direct effects on our conditions of life and on the rich nature of our region, with its three national and five regional nature parks. They have also assimilated publicity about the ongoing evolution of life. Development of ever-larger areas, deforestation, homogenization of the biota by successive waves of invasive introduced species, not to mention global impacts of changing climate induced by the greenhouse effect—all these have become better and better known as menaces to the life that surrounds us. These people are all aware that these scourges are on the increase. This is why a consensus has been reached on each subject we have analyzed, on every budgetary line we have examined: they all want the most effective use of available funds to limit this damage and to convince the political authorities to agree with us. The key part of the budget is therefore destined for better waste treatment and water purification. The rare points of disagreement have generated lively debate on problems of conserving habitat or species that might get in the way of human economic development.

In fact, the members of my commission are also part of 11 other commissions of the same regional institution, commissions with more humanitarian main concerns. These commissions especially question any restrictions on economic development and any exhaustion of natural resources vital to the economy or the lives of the inhabitants. The members of these commissions are therefore especially sensitive to the materialistic and egoistic strategies of our species: support of development (which must be sustainable), economic competition (global or not), improvement of the conditions of life for all individuals (social actions, solidarity among people, the reinforcement of human rights), and lengthening of life (we must live as long as possible in good health).

It is in this context that I have confronted the material realities of the management of nature and mankind. I have observed that all of my science and all of my knowledge amounts to very little in the face of the dilemma posed to decision-makers. Up to what point can we thwart economic development in order to preserve the environment? To convince in straightforward language, to attempt to sensitize people to the most elementary principles of saving species—these are not easy tasks. But after our working sessions, during lunch, I often brought up the subjects of the reflections that have driven me to write this book. Whenever I did this, I saw looks of admiration and surprise. The members of my commission and the political figures were always glad to hear my discourses on the origin of life and its present fragility. I made them discover life in much the same way as Van Leeuwenhoek did when he displayed his animalcules to the courts of Europe.

From this experience, as well as from my sociological analyses of my colleagues on the economic and social council and of regional politicians, I have become convinced that we must greatly improve environmental education. We must also find in spirituality and in ideologies new support that will allow frank and thorough consideration of the scientific evidence on the history of the geneses of life and its past evolution. Only then can we achieve better understanding of what is at stake in the future; a union of all types of thought will lead humanity to take responsibility for its future, indissoluble from the future of the life that surrounds it.

Instead of sinking into a fatalistic attitude and accepting the deleterious consequences of our overpopulation and its impacts, instead of taking refuge in the hope of a glorious future in heaven, we must make everyone realize his or her responsibility to confront right now the great ecological dangers here below.

We must keep, as Van Leeuwenhoek did, our hand firmly resting on the table of material realities of the world, we must affirm that our consciousness of ecological danger will come into play in our everyday lives.

In my functions of teacher, researcher, and environmental manager, I very often hear the same plea: "I know I must do something for our environment and for biodiversity, but how can I be useful at my humble level?"

By the time someone asks such a question, he or she has already accomplished a great deal. In asking that question, he or she has understood or felt the pressure of human impacts on nature, on our environment. He or she is now ready to act. Lung specialists are unanimous: a smoker cannot break the habit unless he or she understands the noxious effect of the smoke on his or her own body and clearly wants to quit.

This first step is therefore indispensable. We have to recognize our growing impact on the life around us. Just as we are often sensitized and revolted by the dismal lives of unfortunate individuals we see or by entire peoples suffering from famine or war (which has helped lead to love of one's neighbor, improvements in human rights, and a more equitable sharing of the wealth of our planet), we must better understand the global problems engendered by erosion of biodiversity, by exhaustion of natural resources, and by the accumulation of pollutants induced by a constantly increasing number of humans.

This sensitization should lead to a second simple and useful step that allows a better comprehension of the silent nature that surrounds us and that we exploit. To this end, we must:

- become conscious of the history and status of the organisms around us. They carry genetic information about our common ancestors. We have the same roots, and together we have evolved on parallel paths since the dawn of time. We should understand that the diversity of life is part of us.
- Marvel at the many strategies for survival and reproduction.
- Become aware of the beauty of life. To know nature better, to feel it better, to become sensitive to all life around us—that is above all what is missing. To redress this lack is the first necessary action.

Everyone at his or her own level can receive an environmental education from various sources. Biologists should not forget the foundations of their science, but they should get better at placing their discoveries in context and try much harder to communicate them, integrated into the knowledge base of their specialty.

Farmers, hunters, and fishers, respectful of the riches they harvest, hikers, animal- or plant-lovers, members of nature or environmental protection organizations, natural science instructors—all these people have much to teach us. They constitute a diverse array of nature experts from whom we can improve our knowledge of nature and our behavior towards it.

To preach increasing sensitization of the public to living nature does not constitute naive proselytization; it is good sense. It should not be done with excessive romanticism, nor in association with fundamentalist "deep ecology." It should begin with ordinary elementary education. Nothing has ameliorated the living conditions of women more than their access to education, and nothing will do as much good for mankind as a better education about everything concerning life, its history, its mechanisms, its fragility. The beauty of life should also be taught just as is the beauty of artistic creations. Whoever knows, admires, and understands life better will love it, respect it, and protect it more. There are no dogmas, rites, or traditions for acquiring and transmitting this knowledge. Whatever the religion or the philosophical stance, every current of thought should, in its own way, contribute to sensitizing us to knowledge about life and to all the problems of managing life and its environment. We must take charge of the biodiversity of life that is now threatened. Only the application of one of the historical strategies of life—unity is strength—can allow us to achieve this objective.

In this context, everyone to his own ability—each a simple ant among many—can make an effort in this direction with the determination of the princes of the House of Orange-Nassau: "It is not necessary to have hope in order to try, nor to succeed in order to persevere."[7] The French motto of the kingdom in which Vermeer lived—*Je maintiendrai* (I will stick to my position)—is, for this objective, a healthy determination.

My modest contribution should be taken in this sense. The history of the geneses of life and of evolution that you have just read is not a fable or a parable or a metaphysical essay; it is a story that comes close to depicting reality. It is the true path of life that we are on and that I have wanted to share with all the enthusiasm that so moves me.

Unlike the reconstruction of the history of the past where many data on life are forever lost, the situation tomorrow will be well known and we will live in it, for better or worse. What will life be like tomorrow? What will nature be in the coming centuries?

A tiny bit at the end of the string of evolution, but crucial for the future of mankind . . .

That is another story!

NOTES

Chapter 1

1. Its discoverers wrote the first work on the Chauvet Cave. The English edition emphasizes the ancestral priority of the paintings (Chauvet J.-M., E. Brunel-Deschamps, and C. Hillaire. 1996. *Dawn of Art: The Chauvet Cave: The Oldest Known Paintings in the World* [New York: Harry N. Abrams]). A group of paleontologists has written a more academic book (Clottes J. 2001. *La Grotte Chauvet. L'art des origines.* [Paris: Seuil]).

2. One of the co-owners of the land above the cave has written a book that discusses Henri's entire judicial saga in relation to the cave. Peschier P. 1999. *La grotte pour 25 centimes. La petite histoire de la grotte Chauvet à Vallon-Pont d'Arc en Ardèche* (Valence: E&R/La Bouquinerie).

3. One of the oldest and most complex illustrations of this genesis is seen on a cupola of the Basilica of San Marco in Venice. It is represented in the form of a comic strip (of 26 frames) by mosaics. The paleontologist and biologist Stephen J. Gould, in the last book of his long series on natural history, discussed at length this representation from the thirteenth century. Gould S.J. 2002. *I Have Landed: The End of a Beginning in Natural History* (New York: Harmony Books; see especially chapter 20).

4. Stephen J. Gould wrote a historic analysis of various calculations establishing these dates. Gould S.J. 1993. *Eight Little Piggies: Reflections in Natural History* (New York: W.W. Norton; see chapter 12).

5. Gould S.J. 1989. *Wonderful Life: The Burgess Shale and the Nature of History* (New York: W.W. Norton).

6. The skeleton of the first placental mammal, *Eomaia scansoria,* was discovered in China. Ji Q., Z.-X. Luo, C.-X. Yuan, J.R. Wible, J.-P. Zhang, and J.A. Georg. 2002. "The earliest known eutherian mammal," *Nature* 416: 816–22.

7. Tattersall I., and J.H. Matternes. 2000. "Once we were not alone," *Scientific American* 182: 56–62.

8. Lucy's 3.5–3.18 million-year-old skeleton was discovered in 1974 by the paleoanthropologists Donald Johanson and his student Tom Gray in the Hadar region of Ethiopia (two Frenchmen, Yves Coppens and Maurice Taieb, participated in this expedition).

Johanson D.C., and M. Taieb. 1976. "Plio-Pleistocene hominid discoveries in Hadar, Ethiopa," *Nature* 260: 293–97; Johanson D., and M. Edey, 1981. *Lucy: The Beginnings of Human Evolution* (New York: Simon & Schuster).

9. The oldest vestige of humanity was uncovered in 2001 by a French and Chadian team in Africa in the Sahel at Toros-Menalla in Chad. The bones are seven million years old and belong to a new hominid species, *Sahelanthropus tchadensis,* nicknamed "Toumaï." Brunet M., F. Guy, D. Pilbeam, et al. 2002. "A new hominid from the Upper Miocene of Chad, Central Africa," *Nature* 418: 145–51; Brunet M., F. Guy, D. Pilbeam, et al. 2005. "New material of the earliest hominid from the Upper Miocene of Chad," *Nature* 434: 752–55; Zollikofer C.P.E., M.S. Ponce de Leon, D.E. Lieberman, et al. 2005. "Virtual reconstruction of *Sahelanthropus tchadensis,*" *Nature* 434: 755–59.

10. Langaney, A., J. Clottes, J Guilaine, and D. Simonet. 1998. *La plus belle histoire de l'Homme* (Paris: Seuil).

11. In 1975, the geneticist Mary-Claire King and the biochemist Allan Wilson established that 99 percent of the proteins and nucleic acid sequences of humans and chimpanzees are identical (King M.-C., and A. Wilson. 1975. "Evolution at two levels in humans and chimpanzees," *Science* 188: 107–16). In 2005, the genome of the chimpanzee was sequenced; there are differences from the human genome in only 35 million of the more than three billion nucleotides (1.23%). (Chimpanzee Sequencing and Analysis Consortium. 2005. "Initial sequence of the chimpanzee genome and comparison with the human genome," *Nature* 437: 69–87; Li W.H., and M.A. Saunders. 2005. "The chimpanzee and us," *Nature* 437: 50–51).

12. In fact, we owe this then-heretical idea to the astronomer Nicolaus Copernicus (1473–1543). His book *De revolutionibus,* in which this theory is explained, was published just before his death.

13. Two martyrs of opponents of deviation from the doctrines of Genesis are especially well known. Giordano Bruno (1548–1600), a Dominican friar, was burned alive in Rome by Pope Clement VIII, and Lucilio Vanini (1585–1619), an Italian priest, had his tongue cut out, was strangled, and then finally burned at Toulouse by religious fanatics. More writers risked religious censure and survived, including Benoit de Maillet (1656–1738), Bernard Le Bouyer de Fontenelle (1657–1757), Montesquieu (Charles Louis de Secondat, baron de la Brede et de Montesquieu,1689–1755), Voltaire (François Marie Arouet, 1694–1778), and Denis Diderot (1713–1784), among others. Several of their writings were censored and some of the authors were jailed for ideas that did not conform to those of the clergy. Closer to our times, in the years 1939–1965, the sinister Stalinist biologist Trofim D. Lysenko (1898–1976) sent to the gulag and death several dozen of his colleagues who contested his new theory of evolution of species by rapid changes (from one species to another) induced by climatic change.

14. Langaney A. 1999. *La philosophie . . . biologique* (Paris: Belin); Lamarck J.-B. de. 1800. *Discours d'ouverture du cours de l'An VIII.* This text is a lecture that was first published in 1801 in the following, more comprehensive work: Lamarck J.-B. de. 1801. *Système des animaux sans vertèbres, ou tableau général des classes, ordres et des genres de ces animaux, présentant leurs caractères essentiels et leur distribution d'après les considérations de leurs rapports naturels et de leurs organisation et suivant l'arrangement établi dans les galeries du Muséum d'Histoire naturelle parmi leurs dépouilles conservées; précédé du discours d'ouverture de l'an VIII de la République, le 21 floréal* (Paris: l'auteur et Déterville). See also Lamarck J.B. de. 1809. *La philosophie zoologique* (Paris: Dentu).

15. David Quammen, the American essayist, has engagingly recounted the story of the paternity of the discovery of the theory of natural selection and of the ambiguous

relationship between Wallace and Darwin: Quammen D. 1996. *The Song of the Dodo: Island Biogeography in an Age of Extinctions* (New York: Simon Schuster).

16. Darwin C. 1859. *On the Origin of Species,* ed. G. Suriano (rev. ed., New York: Grammercy, 1998).

17. Darwin C. 1871. *The Descent of Man and Selection in Relation to Sex* (London: John Murray; rev. ed., Princeton: Princeton University Press, 1981).

18. The Dominican monk who foreswore his writings was Dalmas Leroy (1846–?). Leroy D. 1891. *L'evolution restreinte aux espèces organiques* (Paris: Delhomme et Briguet).

19. The French philosopher Dominique Lecourt has written an excellent book recapping the problems Darwin encountered and the creationist controversies in the United States in the twentieth century (*L'Amérique entre la Bible et Darwin* [Paris: Presse Universitaire de France, Science histoire et Société, 1992]).

20. Leakey R.E. 1994. *The Origin of Humankind* (London: Orion); Leakey R.E., and R. Lewin, 1977. *Origins: What New Discoveries Reveal about the Emergence of our Species and Its Possible Future* (London: Rainbird; New York: E.P. Dutton).

21. In light of modern knowledge of genetics, I have summarized Darwin's work on the interwoven evolution of memory and written culture as it appears in *The Descent of Man and Selection in Relation to Sex.*

22. There are many books on the emergence of self-consciousness. One recent one that caught my attention is Denton D. 1993 *The Pinnacle of Life: Consciousness and Self-Awareness in Humans and Animals* (St. Leonards, Australia: Allen and Unwin).

23. Stephen J. Gould has written on artists' difficulties representing biblical scenes in *The Flamingo's Smile: Reflections in Natural History* (New York: W. W. Norton, 1985; see especially chapter 6). However, it is noteworthy that many artists of the Middle Ages (a period when biblical subjects dominated pictorial art) drew beautiful navels on Adam and Eve; for example, this was done in the eight magnificent depictions of Adam and Eve by Lucas Cranach the Elder (1472–1553). Thus, I like to look for this detail when I visit great art museums.

24. King and Wilson, "Evolution at two levels in humans and chimpanzees."

25. Pierre Teilhard de Chardin was a Jesuit priest and philosopher who taught both theology and paleontology. He constructed a theory of evolution while trying to reconcile has faith with his scientific knowledge. He denied the dogma of original sin and was therefore censured and harried by the Catholic hierarchy (Teilhard de Chardin P. 1955. *Le phénomène humain* [Paris: Seuil]; English language edition, 1965. *The Phenomenon of Man* [New York: Harper & Row]).

26. Professor Henri de Lumley directed these excavations. He has written many syntheses of human evolution, including *L'Homme premier. Préhistoire, évolution, culture* (Paris: Odile Jacob, 1998).

Chapter 2

1. The Russian biochemist Alexander I. Oparin (1894–1980) was the pioneer alchemist of life. He founded a new science: prebiotic chemistry. His initial work, originally published in 1924, has been frequently reprinted in various editions (Oparin A.I. 1957. *The Origin of Life on Earth* [New York: Academic Press]).

2. In 1953 a young doctoral student (then 23 years old), Stanley L. Miller (1930–2007), had the audacity and imagination to test a machine designed to create life. For this historic experiment, he simulated primitive lightning with a device that generated electric arcs in a chamber filled with a gaseous mixture composed of methane, ammonia, and hydrogen,

communicating with another chamber containing water brought to the boiling point. The water evaporated and condensed in the chamber containing the gas. After several weeks, Miller obtained amino acids, molecules typically produced by living organisms. But the choice and concentrations of the gaseous ingredients used in these experiments did not match the composition now believed to have characterized our primitive atmosphere (Miller S.L. 1953. "A production of amino acids under possible primitive earth conditions," *Science* 117: 528–29).

3. In 1976, the evolutionist Richard Dawkins, professor at Oxford University, elaborated a theory of spontaneous molecular generation from a self-reproducing primer, which he called a replicator. He described the phases of this process in the minutest detail. The theory is hypothetical, not based on empirical evidence (Dawkins R. 1976. *The Selfish Gene* [Oxford: Oxford University Press]). This idea has been taken up by other scientists, including the Nobel laureates Walter Gilbert, originator of the concept of the "RNA world," and Christian de Duve, who published a synthesis of technical and theoretical advances on this theme (Gilbert W. 1986. "The RNA world," *Nature* 319: 618; Duve, C. de. 2005. *Singularities: Landmarks on the Pathways of Life* [Cambridge: Cambridge University Press]).

4. Several research teams have obtained results on the selection of good molecular shapes for the construction of RNA using inorganic substances as a matrix (notably clay such as montmorillonite). A German biologist, Günter Wächtershäuser, began these attempts (Wächtershäuser G. 1988. "Before enzymes and templates: theory of surface metabolism," *Microbiological Reviews* 52: 452–84; idem, 1990. "Evolution of the first metabolic cycles," *Proceedings of the National Academy of Sciences U S A* 87: 200–4; Hazen R.M., T.R. Filley, and G.A. Goodfriend. 2001. "Selective adsorption of L- and D-amino acids on calcite: implications for biochemical homochirality," *Proceedings of the National Academy of Sciences U S A* 98: 5487–90; Ertem G., and J.P. Ferris. 1996. "Synthesis of RNA oligomers on heterogeneous templates," *Nature* 379: 238–40; Ferris J.P., A.R. Hill Jr., R. Liu, and L.E. Orgel. 1996. "Synthesis of long prebiotic oligomers on mineral surfaces," *Nature* 381: 59–61; Ertem G., and J.P. Ferris. 1998. "Formation of RNA oligomers on montmorillonite: site of catalysis," *Origin of Life and Evolution of the Biosphere* 28: 485–99).

5. Alexander G. Cairns-Smith first hypothesized the formation of DNA in a clay matrix (*Genetic Takeover and the Mineral Origins of Life* [Cambridge: Cambridge University Press, 1982]; *Seven Clues to the Origin of Life: A Scientific Detective Story* [Cambridge: Canto, Cambridge University Press, 1986]). According to him, the crystalline organization of kaolinite would have been able to serve as a template for the primer of DNA. However, Cairns-Smith did not exclude the hypothesis of panspermia. (See also Cairns-Smith A.G., and H. Hartman. 1986. *Clay Minerals and the Origin of Life* [Cambridge, Cambridge University Press]). Leslie E. Orgel (1927–2007) proposed hypotheses on the synthesis of life from molecules simpler than DNA (that is, RNA) that polymerized spontaneously in rocky interstices. These molecules would have catalyzed the formation of the precursors of life (Joyce G.F., and L.E. Orgel. 1993. "Prospects for understanding the origin of the RNA world," in R.F. Gesteland and J.F. Atkins, eds. *The RNA World,* [Cold Spring Harbor, NY: Cold Spring Harbor Laboratory Press], 1–25). See also Liu R., and L.E. Orgel. 1998. "Polymerization on the rocks: amino acids and arginine," *Origins of Life and Evolution of the Biosphere* 28: 245–57. Orgel collaborated with the Nobel laureate Francis Crick to suggest the arrival of life on Earth in a spaceship that crashed on our planet 3.9 billion years ago.

6. Genesis 2:7 (New International Version): "The Lord God formed the man from the dust of the ground and breathed into his nostrils the breath of life, and man became a living being."

7. The expression *primordial soup* was coined in 1929 by the British geneticist John B.S. Haldane (1892–1964).

8. By 1922, Alexander I. Oparin had succeeded in creating small spherical structures called *coacervates* from an emulsion containing proteins and gum arabic. These spheres in suspension measured several microns in diameter. Since then, starting without precursors from living organisms, several teams have synthesized vesicles analogous to cell membranes. The Belgian biochemist and Nobel laureate in medicine Christian de Duve is responsible for significant breakthroughs in our knowledge of the structural and functional organization of the living cell. In a well-researched work, he described all the hypotheses about the formation of cell walls and of the first genes, all the while recognizing that they all lead to blind alleys. None has been verified to this day. He hypothesized an extraordinary stroke of luck as having led, quite suddenly, to the emergence of a molecule able to self-replicate and to guide the construction of other molecules (Duve C. de. 1995. *Vital Dust: Life as a Cosmic Imperative* [New York: Basic Books]). Finally, it is important to note the hypothesis of Jack Szostak's team, associating a pinch of RNA and self-reproducing vesicules in a formation called *protocells*. This synthesis is catalyzed by a type of clay, montmorillonite. (Szostak J.W., D.P. Bartel, and P.L. Luisi. 2001. "Synthesizing life," *Nature* 409: 387–90; Hanczyc M.M., S.M. Fujikawa, and J.W. Szostak. 2003. "Experimental models of primitive cellular compartments: encapsulation, growth and division," *Science* 302:618–22).

9. Christian de Duve has published three popular books summarizing all research relevant to mechanisms that could have led to life (*Singularities*; *Life Evolving: Molecules, Mind and Meanings* [New York: Oxford University Press, 2002]; and *Vital Dust)*.

10. Dalton R. 2002. "Squaring up over ancient life," *Nature* 417: 782–84.

11. Mojzsis S.J., G. Arrhenius, K.D. McKeegan, et al. 1996. "Evidence for life on earth before 3,800 million years ago," *Nature* 384: 55–58.

12. In 2002, two research teams proposed a chemical explanation for the formation of organic molecules discovered in rocks 3.8 billion years old (Van Zuilen M.A., A. Lepland, and G. Arrhenius. 2002. "Reassessing the evidence for the earliest traces of life," *Nature* 418: 627–30; Fedo C.M., and M.J. Whitehouse. 2002. "Metasomatic origin of quartz-pyroxene rock, Akilia, Greenland, and implications for earth's earliest life," *Science* 296: 1448–52). These papers were then debated: Mojzsis S.J., and T.M. Harrison. 2002. "Origin and significance of Archean quartzose rocks at Akilia, Greenland," *Science* 298: 917; Friend C.R.L., A.P. Nutman, and V.C. Bennet. 2002. "Origin and significance of Archean quartzose rocks at Akilia, Greenland," *Science* 298: 917; Fedo, C.M., and M.J. Whitehouse, 2002. "Origin and significance of Archean quartzose rocks at Akilia, Greenland," *Science* 298: 917; Palin J.M., C.M. Fedo, and M.J. Whitehouse. 2002. "The origin of a most contentious rock," *Science* 298: 961–62.

13. J. William Schopf, professor at the University of California, Los Angeles, specializes in research on the oldest fossils. In the course of his work in Australia, he discovered the oldest traces of life (Schopf J.W. 1993. "Microfossils of the early Archean apex chert: new evidence of the antiquity of life," *Science* 260: 640–46; idem, 2002. "When did life begin?" in Schopf J.W., ed. *Life's Origin: The Beginnings of Biological Evolution* [Berkeley: University of California Press, 2002], 158–79); idem, 2006. "Fossil evidence of Archaean life," *Philosophical Transactions of the Royal Society* B 361: 869–85). Other teams have dated microfossils from South Africa and Australia to between 3.5 and 3.45 billion years old (Knoll A.H., and E.S. Barghoorn. 1977. "Archean microfossils showing cell division from the Swaziland system of South Africa," *Science* 198: 396–98; Walsh M.M., and D.R. Lowe. 1985.

"Filamentous microfossils from the 3,500-Myr-old Onverwacht Group, Barberton Mountain Land, South Africa," *Nature* 314: 530–32; Walsh M.M. 1992. "Microfossils and possible microfossils from the early Archean Onverwacht Group, Barberton Mountain Land, South Africa," *Precambrian Research* 54: 271–93; Buick R. 1990. "Microfossil recognition in Archean rocks: an appraisal of spheroids and filaments from a 3500 M.Y. old chert-barite unit at North Pole, Western Australia," *Palaios* 5: 441–59).

14. In 2002, the controversy surrounding the oldest fossils described by J. William Schopf was very heated (Brasier D.M., O.R. Green, A.P. Jephcoat, et al. 2002. "Questioning the evidence for earth's oldest fossils," *Nature* 416: 76–81; Pasteris J.D., and B. Wopenka. 2002. "Images of the earth's earliest fossils?" *Nature* 420: 476–77). J. William Schopf and his colleagues responded to these criticisms (Schopf J.W., A.B. Kudryavtsev, D.G. Agresti, T.J. Wdowiak, and A.D. Czaja. 2002. "Images of the earth's earliest fossils?" *Nature* 420: 477). Other teams have pointed to aspects of the controversy that tend to show that the filamentous forms described by Schopf could have been of inorganic origin (Garcia-Ruiz J.M., S.T. Hyde, A.M. Carnerup, et al. 2003. "Self-assembled silica-carbonate structures and detection of ancient microfossils," *Science* 302: 1194–97; De Gregorio B.T., and T.G. Sharp. 2003. "Determining the biogenicity of microfossils in the Apex chert, Western Australia, using transmission electron microscopy," *Lunar and Planetary Science* 34: 1267–68).

15. *Stromatolite* comes from *stromato* (meaning carpet) and *lith* (meaning rock). The first stromatolites, found by the American paleontologist Charles Doolittle Walcott at the end of the nineteenth century, were called Cryptozoons.

16. Lowe D.R. 1994. "Abiological origin of described stromatolites older than 3.2 Ga," *Geology* 22: 387–90.

17. Hofmann H.J. 2000. "Archean stromatolites as microbial archives," in Riding R., and S.W. Awramik, eds. *Microbial Sediments* (Berlin and Heidelberg: Springer-Verlag), 315–27; Schopf, "Fossil evidence of Archaean life."

18. Hofmann H.J., K.Grey, A.H. Hickman, and R.I. Thorpe. 1999. "Origin of 3.45 Ga coniform stromatolites in Warrawoona group, Western Australia," *Geological Society of America Bulletin* 111: 1256–62.

19. Allwood A., M.R. Walter, B.S. Kamber, C.P. Marshall, and I.W. Birch. 2006. "Stromatolite reef from the early Archean era of Australia," *Nature* 441: 714–18.

20. Burns B.P., F. Goh, M. Allen, and B.A. Neilan. 2004. "Microbial diversity of extant stromatolites in the hypersaline marine environment of Shark Bay, Australia," *Environmental Microbiology* 6: 1096–1101.

21. Stromatolites growing at substantial depths were discovered in 1983 in the Atlantic Ocean, near the Bahamas. (Dravis J.J. 1983. "Hardened subtidal stromatolites. Bahamas," *Science* 219: 385–86; Dill R.F., E.A. Shinn, A.T. Jones, K. Kelly, and R.P. Steinen. 1986. "Giant subtidal stromatolites forming in normal salinity waters," *Nature* 324: 55–58).

22. Several syntheses have been published on the chemical traces left by the first bacteria. These discuss oxidized minerals, deposits of carbon with low concentrations of the isotope ^{13}C (probably traces of photosynthesis by plants), and accumulations of sulfates (in which the percentages of different isotopes of sulfur also suggest activity by plant-like bacteria). See Knoll A.K. 2003. *Life on a Young Planet: The First Three Billion Years of Evolution on Earth* (Princeton, NJ: Princeton University Press; see especially chapter 6); Copley J. 2001. "The story of O," *Nature* 410: 862–64. Hiroshi Ohmoto is the most committed advocate of the view that oxygen appeared very early on Earth (more than 3.5 billion years ago) and was produced by plant-like bacteria. See Ohmoto H. 1996. "Evidence in pre-2-2

Ga paleosols for the early evolution of atmospheric oxygen and terrestrial biotas," *Geology* 24: 1135–38.

23. J. William Schopf used the metaphor of the Volkswagen syndrome in his book *Cradle of Life*. In the same book, he described the 11 species of bacteria found in the oldest stromatolites, some of which have the appearance of multicellular filaments similar to present-day cyanobacteria. Schopf also stressed the abundance of discoveries of very ancient (Precambrian) stromatolites spread over several continents (250 sites in 12 nations). See Schopf J.W. 1999. *Cradle of Life: The Discovery of Earth's Oldest Living Fossils* (Princeton, NJ: Princeton University Press).

24. Rasmussen, B. 2000. "Filamentous microfossils in a 3,235 million year old volcanogenic massive sulphide deposit," *Nature* 405: 676–79.

25. Furnes H., N.R. Banerjee, K. Muehlenbachs, H. Staudigel, and M. de Wit. 2004. "Early life recorded in archean pillow lavas," *Science* 304: 578–81.

26. Shen Y., R. Buick, and D.E. Canfield. 2001. "Isotopic evidence for microbial sulphate reduction in the early Archaean era," *Nature* 401: 77–81.

27. Peter Ward developed this theme while integrating viruses into systems for the classification of life (Ward P. 2005. *Life as We Do Not Know It: The NASA Search for (and Synthesis of) Alien Life* [London: Viking Penguin]).

28. A Finnish research team has provided the most precise description of nanobacteria (Kajander E.O., and N. Çiftçioğlu. 1998. "Nanobacteria: an alternative mechanism for pathogenic intra- and extracellular calcification and stone formation," *Proceedings of the National Academy of Sciences U S A* 95: 8274–79). These authors also presented a synthesis of their results during a meeting of the leading specialists on the minimum size of living organisms (Kajander, E.O., M. Björklund, and N. Çiftçioğlu. 1999. "Suggestions from observations on nanobacteria isolated from blood," in Space Studies Board. Commission on Physical Sciences, Mathematics, and Applications. *Size Limits of Very Small Microorganisms, Proceedings of a Workshop* [Washington, DC: National Academy Press], 50–55). Aspects of these conclusions have recently been questioned in Cisar J.O., D.-Q. Xu, J. Thompson, W. Swaim, L. Hu, and D.J. Kopecko. 2000. "An alternative interpretation of nanobacteria-induced biomineralization," *Proceedings of the National Academy of Sciences U S A* 97: 11511–15.

29. Steering group for the workshop on size limits of very small microorganisms. Space Studies Board. Commission on Physical Sciences, Mathematics, and Applications. National Research Council, *Size Limits of Very Small Microorganisms*. Subsequent to this theoretical conclusion, even tinier forms of life have been discovered. Some of these microbes, which belong to the Archaea and are termed Archaeal Richmond Mine Acidophilic Nanoorganisms (ARMAN), are about 200 nanometers (0.2 microns) long and have a volume of 0.006 cubic micrometers (Baker B.J., G.W. Tyson, R.I. Webb, J. Flanagan, P. Hugenholtz, E.E. Allen, and J.F. Banfield. 2006. "Lineages of acidophilic Archaea revealed by community genomic analysis," *Science* 314: 1933–35). The tiniest microorganisms belonging to the phylum of bacteria were found trapped in ice for the last 120,000 years and are about 0.011 to 0.07 cubic micrometers. They passed through 0.2-micron filters (Miteva V.I., P.P. Sheridan, and J.E. Brenchley. 2005. "Phylogenetic and physiological diversity of microorganisms isolated from a deep Greenland ice core," *Applied and Environmental Microbiology* 71: 202–13). Miteva, V.I., and J.E. Brenchley. 2005. "Detection and isolation of ultrasmall microorganisms from a 120,000-year-old Greenland glacier ice core," *Applied and Environmental Microbiology* 71: 7806–18.

30. Hutchinson C.A. III, S.N. Peterson, S.R. Gill, R.T. Cline, O. White, C.M. Fraser, H.O. Smith, and J.C. Venter. 1999. "Global transposon mutagenesis and a minimal

Mycoplasma genome," *Science* 286: 2165–69; Gill R., F.J. Silva, J. Peretó, and A. Moya. 2004. "Determination of the core of a minimal bacterial gene set," *Microbiology and Molecular Biology Reviews* 68: 518–37.

31. The term LUCA was proposed in 1996 at a colloquium not far from Nice, France. It joined the term *progenote,* coined by Carl Woese and George Fox in 1977 to describe primitive prebacterial stages, and *Cenancestor,* suggested by Walter Fitch and Kim Upper in 1987 to cover an ancestral ensemble that led to bacteria (from the Greek *Cen,* meaning ensemble). See Woese C.R., and G.E. Fox. 1977. "The concept of cellular evolution," *Journal of Molecular Evolution* 10: 1–6; Fitch W.M., and K. Upper. 1987. "The phylogeny of tRNA sequences provides evidence for ambiguity reduction in the origin of the genetic code," *Cold Spring Harbor Symposia on Quantitative Biology* 52: 759–67; Lazcano A., and P. Forterre. 1999. "The molecular search for the last common ancestor," *Molecular Evolution* 49: 411–12.

32. Carl Woese emphasizes the fact that "the universal ancestor is not an entity, not a thing. It is a process characteristic of a particular evolutionary stage" (Woese C.R. 1998. "The universal ancestor," *Proceedings of the National Academy of Sciences U S A* 95: 6854–69; quote from p. 6858).

33. In 1908, the Swedish Nobel laureate in physics, Svante Arrhenius (1859–1927), coined the term *panspermia* for his theory, according to which the seeds of life existed somewhere in space and colonized Earth. But the idea of life dispersed throughout the universe goes back to the Greek philosopher Anaxagoras of Klazomenai (ca. 500/497 BC to ca. 428 BC, who was impressed by a meteorite impact at Aigospotamos around 469 BC). At the end of the nineteenth century, two authors proposed that bacteria were transported by meteorites: the German L.F. von Helmholtz (1821–1894) and the Irish Nobel laureate, William Thomson (1824–1907), honored by the title of Lord Kelvin.

34. Crick F.C. 1981. *Life Itself: Its Origin and Nature* (New York: Simon & Schuster).

35. Through a fortuitous professional encounter, I met Professor Daniel Simberloff, an ecologist and specialist in conservation biology. Simberloff honored me by translating into English my previous book on the history of the invasion of the "killer alga" *Caulerpa taxifolia,* which has spread in the Mediterranean since 1984. As it happens, for many years, Friedmann's and Simberloff's offices were across the hall from one another at Florida State University, a fact that helped Friedmann to follow my research on and problems related to the discovery of this algal invasion.

36. The proportions of the pairs uranium 238/lead 206 and rubidium 87/strontium 87 are most often used to date rocks and meteorites older than 3 billion years. The higher the proportion of stable atoms (lead and strontium), the older the rock.

37. Carbon in nature is composed of the following: 99 percent of the stable isotope ^{12}C, 1 percent of the relatively stable ^{13}C, and an infinitesimal quantity (a bit more than one part per trillion) of the unstable (radioactive) isotope ^{14}C. This last isotope is used for dating "recent" substances, such as charcoal from caves inhabited by Cro-Magnons; after a few dozen millennia, this isotope is so rare that it cannot be measured reliably.

38. The oldest trace evidence of the first continents of our planet dates to 4.4 billion years ago—a tiny bit of zircon found in Australia. Other traces, more firmly established (by many samples), are from 3.9 billion years ago (rocks from Greenland and northwest Canada). The "oldest" meteorite known today dates from 4.56 billion years ago. When it fell to Earth in 1969 in Allende, Mexico, it broke into several fragments with a total weight of two tons. This rock contains carbon and crystals from another star. Slices of it have been sold on the Internet for over $200.

39. McKay D.S., E.K. Gibson Jr., K.L. Thomas-Keprta, H. Vali, C.S. Romanek, S.J. Clemett, X.D.F. Chillier, C.R. Maechling, and R.N. Zare. 1996. "Search for past life on

Mars: Possible relic biogenic activity in Martian meteorite ALH84001," *Science* 273: 924–30. A scientific reporter, Kathy Sawyer, has detailed the discovery of the meteorite by Robbie Score and the story of the engagement of David McKay's scientific team in research on the meteorite (Sawyer K. 2006. *The Rock from Mars: A Detective Story on Two Planets* [New York: Random House]).

40. A press release titled "Meteorite Yields Evidence of Primitive Life on Early Mars" was distributed for NASA on August 7, 1996, by Donald L. Savage (NASA headquarters in Washington), James Hartsfield (Johnson Space Center in Houston, Texas), and David Salisbury (Stanford University in Palo Alto, California).

41. Thomas-Keprta K.L., D.S. McKay, S.J. Wentworth, T.O. Stevens, A.E. Taunton, C.A. Allen, A. Coleman, E.K. Gibson Jr., and C.S. Romanek. 1998. "Bacterial mineralization patterns in basaltic aquifers: implications for possible life in Martian meteorite ALH84001," *Geology* 26: 1031–35.

42. Bradley J.P., R.P. Harvey, and H.Y. McSween Jr. 1997. "No nanofossils in Martian meteorite," *Nature* 390: 454–55; McKay D.S., E.K. Gibson Jr., K.L. Thomas-Keprta, and H. Vali. 1997. "No nanofossils in Martian meteorite: reply," *Nature* 390: 455–56.

43. Zolotov M.Y., and E.L. Shock. 2000. "An abiotic origin for hydrocarbons in the Allan Hills 84001 Martian meteorite through cooling of magmatic and impact-generated gases," *Meteoritics and Planetary Science* 35: 629–38; McSween H.Y. Jr., and R.P. Harvey. 1998. "An evaporation model for formation of carbonates in the ALH84001 Martian meteorite," *International Geology Review* 40: 774–83.

44. Steele A., D.T. Goddard, D. Stapleton, J.K.W. Toporski, V. Peters, V. Bassinger, G. Sharples, D.D. Wynn-Williams, and D.S. McKay. 2000. "Investigations into an unknown organism on the Martian meteorite Allan Hills 84001," *Meteoritics and Planetary Science* 35: 237–41.

45. Thomas-Keprta K.L., S.J. Wentworth, D.S. McKay, D. Bazylinski, M.S. Bell, C.S. Romanek, and E.K. Gibson. 1999. "On the origins of magnetite in Martian meteorite ALH84001," *Lunar Planetary Science* 30: abstract 1856; Thomas-Keprta K.L., D.A. Bazylinski, J.L. Kirschvink, S.J. Clemett, D.S. McKay, S.J. Wentworth, H. Vali, E.K. Gibson Jr., and C.S. Romanek. 2000. "Elongated prismatic magnetite crystals in ALH84001 carbonate globules: potential Martian magnetofossils," *Geochimica et Cosmochimica Acta* 64: 4049–81.

46. Magnetite chains, or, magnetosomes, have been found in bacteria since 1975 (Blakemore R. 1975. "Magnetotactic bacteria," *Science* 190: 377–79). Since then, several dozen publications describe these inclusions. Stephen J. Gould devoted a chapter in one of his many books to an explanation of the amazing properties of bacteria equipped with these tiny magnets. He did not know, at that time, that many magnetite chains were going to be found in meteorites (Gould S.J. 1980. *The Panda's Thumb: More Reflections in Natural History* [New York: W. W. Norton; Harmondsworth: Penguin]; see chapter 30). See also these two syntheses on magnetite chains: Schüler D., and R.B. Frankel. 1999. "Bacterial magnetosomes: microbiology, biomineralization and biotechnological applications," *Applied Microbiology* 52: 464–73; Spring S., and K.H. Schleifer. 1995. "Diversity of magnetotactic bacteria," *Systematic and Applied Microbiology* 18: 147–53.

47. Kirschvink J.L., and H. Vali. 1999. "Criteria for the identification of bacterial magnetofossils on Earth or Mars," *Lunar Planetary Science* 30: abstract 1681; Thomas-Keprta et al., "Elongated prismatic magnetite crystals in ALH84001 carbonate globules: potential Martian magnetofossils."

48. Friedmann E.I., J. Wierzchos, and C. Ascaso. 1998. "Chains of magnetite crystals in the meteorite ALH84001," in *Workshop on the issue of Martian meteorites* (Lunar Planetary Institute, Houston, TX), contribution 956: 14–16.

49. Friedmann E.I., J. Wierzchos, C. Ascaso, and M. Winklhofer. 2001. "Chains of magnetite crystals in the meteorite ALH84001: evidence of biological origin," *Proceedings of the National Academy of Sciences U S A* 98: 2176–81.

50. Thomas-Keprta K.L., S.J. Clemett, D.A. Bazylinski, J.L. Kirschvink, D.S. McKay, S.J. Wentworth, H. Vali, E.K. Gibson Jr., M.F. McKay, and C.S. Romanek. 2001. "Truncated hexa-octahedral magnetite crystals in ALH84001: presumptive biosignatures," *Proceedings of the National Academy of Sciences U S A* 98: 2164–69. This discovery was challenged nine months later, creating a lively controversy over crystallization of magnetites by organic or inorganic processes (Buseck P.R., R.E. Dunin-Borkowski, B. Devouard, R.B. Frankel, M.R. McCartney, P.A. Midgley, M. Pósfai, and M. Weyland. 2001. "Magnetite morphology and life on Mars," *Proceedings of the National Academy of Sciences U S A* 98: 13490–95; Thomas-Keprta K.L., S. Clemett, C. Romanek, D.A. Bazylinski, J.L. Kirschvink, D.S. McKay, S.J. Wentworth, H. Vali, and E.K. Gibson Jr. 2002. "Multiple origins of magnetite crystals in ALH84001 carbonates," *Lunar and Planetary Science* 33: abstract 1911; Thomas-Keprta K.L., S.J. Clemett, D.A. Bazylinski, J.L. Kirschvink, D.S. McKay, S.J. Wentworth, H. Vali, E.K. Gibson Jr. and C.S. Romanek. 2002. "Magnetofossils from ancient Mars: a robust biosignature in the Martian meteorite ALH84001," *Applied and Environmental Microbiology* 68: 3663–72).

51. Tan W.C., and S.L. Van Landingham. 1967. "Electron microscopy of biological-like structures in the Orgueil carbonaceous meteorite," *Geophysical Journal of the Royal Astronomical Society* 12: 237.

52. Hoover R.B., A.Y. Rosanov, S.I. Zhmur, and V.M. Gorlenko. 1998. "Further evidence of microfossils in carbonaceous chondrites," *Instruments, Methods and Missions for Astrobiology*. Proceedings of SPIE 3441: 203–16.

53. Vainshtein, M., N. Suzina, and V. Sorokin. 1997. "A new type of magnet-sensitive inclusions in cells of photosynthetic purple bacteria," *Systematic and Applied Microbiology* 20: 182–86.

54. In 1961, George Claus and Bartholomew S. Nagy (1927–1995), followed by the Nobel laureate Harold C. Urey (1893–1981), published a sensational discovery: they had detected fossils of cells in the meteorite Orgueil. They believed the meteorite came from the moon, which had formerly been part of Earth. The moon, according to this hypothesis, had sent a rock with the oldest known fossils. This discovery became the object of heated debates. Eventually, a young French micropaleontologist, Martine Rossignol (who is today a professor in Paris) worked with Friedmann on a fragment of the meteorite and concluded that there were no traces of fossils. As for the chains of bacterial magnetites observed by Vainshtein, they arose from terrestrial contamination (Nagy B., W.G. Meinschein, and D.J. Hennessy. 1961. "Mass spectroscopic analysis of the Orgueil meteorite: evidence for biogenic hydrocarbons," *Annals of the New York Academy of Science* 93: 25–35; Claus G., and B. Nagy. 1961. "A microbiological examination of some carbonaceous chondrites," *Nature* 192: 594–96; Urey H.C. 1962. "Origin of life-like forms in carbonaceous chondrites," *Nature* 193: 1119–23; Urey H.C. 1966. "Biological material in meteorites: a review," *Science* 151: 157; Rossignol-Strick M., and E.S. Barghoorn. 1971. "Extraterrestrial abiogenic organization of organic matter: the hollow spheres of the Orgueil meteorite," *Space Life Sciences* 3: 89–107).

55. These data pertain to four Martian meteorites:

- the meteorite Nakhla, which was found in Egypt and arose on Mars between 1.3 and 0.7 million years ago.
- the meteorite Murchinson, which fell in Australia in 1969.
- the meteorite Efremovka, which fell in Kazakhstan in 1962.
- the meteorite Shergotty, which fell in India in 1865 and arose on Mars between 360 and 165 million years ago.

See Hoover et al. "Further evidence of microfossils in carbonaceous chondrites"; Gibson E.K., D.S. McKay, K.L. Thomas-Keprta, S.J. Wentworth, F. Westall, A. Steele, C.S. Romanek, M.S. Bell, and J. Toporski. 2001. "Life on Mars: evaluation of the evidence within Martian meteorites ALH84001, Nakhla, and Shergotty," *Precambrian Research* 106: 15–34.

56. Another Martian meteorite, which landed in Tatahouine, Tunisia, in 1931, was the object of contradictory research. Fragments were removed and stored in Paris immediately after it landed. Others were taken more recently. Only from these latter fragments have researchers found traces of contamination (carboniferous globules assimilated to nanobacteria) by terrestrial organisms similar to those noticed on other Martian meteorites (Barrat J.A., Ph. Gillet, C. Lécuyer, S. M. F. Sheppard, and M. Lesourd. 1998. "Formation of carbonates in the Tatahouine meteorite," *Science* 280: 412–14; Barrat J.A., Ph. Gillet, M. Lesourd, J. Blichert-Toft, and G.R. Poupeau. 1999. "The Tatahouine diogenite: mineralogical and chemical effects of sixty-three years of terrestrial residence," *Meteorite and Planetary Science* 34: 91–97; Gillet P., J. A. Barrat, T.H. Heulin, W. Achouak, M. Lesourd, F. Guyot, and K. Benzerara. 2000. "Bacteria in the Tatahouine meteorite: nanometric-scale life in rock," *Earth and Planetary Science Letters* 175: 161–67).

57. Spring S., R. Amann, W. Ludwig, K. Schleifer, H. van Gemerden, and N. Petersen. 1993. "Dominating role of an unusual magnetotactic bacterium in the microaerobic zone of a freshwater sediment," *Applied and Environmental Microbiology* 59: 2397–403.

58. Cano R.J., and M. Borucki. 1995. "Revival and identification of bacterial spores in 25 to 40 million year old Dominican amber," *Science* 268: 1060–64.

59. Vreeland R.H., W.D. Rozenzweig, and D.W. Powers. 2000. "Isolation of a 250-million year-old halotolerant bacterium from a primary salt crystal," *Nature* 407: 897–900. These experiments in bacterial regeneration call to mind the plot of *Jurassic Park,* by Michael Crichton, in which insane researchers resuscitate dinosaurs from their blood cells contained in the abdomens of fossilized mosquitoes that bit them more than 65 million years ago.

60. Sykes B. 1997. "Really ancient DNA: lights turning red on amber," *Nature* 386: 764–65; Gutiérrez G., and A. Marín. 1998. "The most ancient DNA recovered from an amber-preserved specimen may not be as ancient as it seems," *Molecular Biology and Evolution* 15: 926–29; Graur D., and T. Pupko. 2001. "The Permian bacterium that isn't," *Molecular Biology and Evolution* 18: 1143–46; Hazen R.M., and E. Roedder. 2001. "How old are the bacteria from the Permian age ?" *Nature* 411: 155.

61. Powers D.W., R.H. Vreeland, and W.D. Rosenzweig. 2001. "How old are the bacteria from the Permian age?" *Nature* 411: 155; Vreeland R.H., and W.D. Rosenzweig, 2002. "The question of uniqueness of ancient bacteria," *Journal of Industrial Microbiology and Biotechnology* 28: 32–41; Maughan H., C.W. Birky Jr., W.L. Nicholson, W.D. Rosenzweig, and R.H. Vreeland. 2002. "The paradox of the 'ancient' bacterium which contains 'modern' protein-coding genes," *Molecular Biology and Evolution* 19: 1637–39.

62. This bacterium was discovered in 1956 by Arthur W. Anderson at an agricultural experiment station in Corvallis, Oregon. He was intrigued by the persistence of bacteria on meat that had been sterilized by a radioactive source. A functional genome has even been

recovered from this bacterium beginning with several hundred fragments produced by intense exposure to radiation (Zahradka K., D. Slade, A. Bailone, S. Sommer, D. Averbeck, M. Petranovic, A.B. Lindner, and M. Radman. 2006. "Reassembly of shattered chromosomes in *Deinococcus radiodurans*," *Nature* 443: 569–73).

63. Venter C.J. 2002. "Whole-genome shotgun sequencing," in M. Yudell and R. DeSalle, eds., *The Genomic Revolution: Unveiling the Unity of Life*. (Washington, DC: Joseph Henry Press), 48–63. A team of researchers from India and Cardiff, Wales, detected "stratospheric tourists": living bacteria collected at an altitude of 41 kilometers! See Harris M.J., N.C. Wickramasinghe, D. Lloyd, J.V. Narlikar, P. Rajaratnam, M.P. Turner, S. Al-Mufti, M.K. Wallis, S. Ramadurai, and F. Hoyle. 2002. "The detection of living cells in stratospheric samples," *Proceedings 46th annual meeting of SPIE* 4495: 192–98.

64. Weiss B.P., J.L. Kirschvink, F.J. Baudenbacher, H. Vali, N.T. Peters, F.A. Macdonald, and J.P. Wikswo. 2000. "A low temperature transfer of ALH84001 from Mars to Earth," *Science* 290: 791–95.

65. Roten C.-A., H. A. Gallusser, G.D. Borruat, S.D. Udry, and D. Karamata. 1998. "Impact resistance of bacteria entrapped in small meteorites," *Bulletin de la Société Vaudoise des Sciences Naturelles* 86: 1–17.

66. Since the publication in January 2001 of two articles on the ALH84001 magnetites by the teams of Friedmann et al. (see note 49 in this chapter) and Thomas-Keprta et al. (note 50), several studies have been undertaken to confirm or refute them. Only the organic origin of all the ALH84001 magnetites has been challenged. In September 2003, a consensus article signed by members of the two teams (including Friedmann and Christopher McKay) was published in the specialist journal *Astrobiology*. The conclusion of this synthesis confirms the possible existence or magnetite crystals of inorganic origin in ALH84001, without excluding the possibility of a biological origin for other crystals (more than 25 percent of the magnetites). However, as concerns the alignment of the crystals (the magnetite necklaces observed by Friedmann) and the regular spacing of the crystals constituting these necklaces, no other explanation than that of biological formation was suggested. The article also lists a number of recommendations to improve the analysis (McKay C.P., I. Friedmann, R.B. Frankel, and D.A. Bazylinski. 2003. "Magnetotactic bacteria on Earth and on Mars," *Astrobiology* 3: 263–70). Another article appeared at the same time, in the same journal, authored by Allan H. Treiman, fierce opponent of all hypotheses claiming a biological origin for elements found in ALH84001. It challenged the article by Thomas-Keprta and colleagues in particular, which argued that all the magnetite crystals from ALH84001 were biological in origin (note 50). In this long article, he briefly mentions the magnetite alignment observed by Friedmann, recognizing that no inorganic mechanism can explain the regularity and sinuous shape of the magnetite necklaces photographed by Friedmann. All the arguments formulated by Friedmann's team in January 2001 therefore remain unchallenged. In 2004, another team confirmed the presence of magnetite chains in ALH84001 (10 percent of the tens of millions of magnetite crystals analyzed may exist in isolated chains) (Weiss B.P., S. Sam Kim, J.L. Kirschvink, R.E. Kopp, M. Sankaran, A. Kobayashi, and A. Komeili. 2004. "Magnetic tests for magnetosome chains in Martian meteorite ALH84001," *Proceedings of the National Academy of Sciences U S A* 101: 8281–84). The magnetite necklaces from ALH84001 still seem very likely to be "skeletons" of exogenous bacteria (Treiman A.H. 2003. "Submicron magnetite grains and carbon compounds in Martian meteorite ALH84001: inorganic, abiotic formation by shock and thermal metamorphism," *Astrobiology* 3: 369–92).

67. Taylor A.P., and J.C. Barry. 2004. "Magnetosomal matrix: ultrafine structure may template biomineralization of magnetosomes," *Journal of Microscopy* 213: 180–97; Schüler

D. 2006. "Magnetoreception and magnetosomes in bacteria," in D. Schüler, ed., *Microbiology Monographs* vol. 3 (Heidelberg, Springer).

68. There are many recent data on both the abundance of liquid water on Mars in the past and the current presence of water as ice. The several publications on this subject include the following: Bibring J.P., Y. Langevin, F. Poulet, A. Gendrin, B. Gondet, M. Berthé, A. Soufflot, P. Drossart, M. Combes, G. Bellucci, V. Moroz, N. Mangold, B. Schmitt, and the OMEGA Team. 2004. "Perennial water ice identified in the south polar cap of Mars," *Nature* 428: 627–30; Haskin L.A., A. Wang, B.L. Jolliff, H.Y. McSween, B.C. Clark, D.J. Des Marais, S.M. McLennan, N.J. Tosca, J.A. Hurowitz, J.D. Farmer, A. Yen, S.W. Squyres, R.A. Arvidson, G. Klingelhöfer, C. Schröder, P.A. de Souza Jr., D.W. Ming, R. Gellert, J. Zipfel, J. Brückner, J.F. Bell III, K. Herkenhoff, P.R. Christensen, S. Ruff, D. Blaney, S. Gorevan, N.A. Cabrol, L. Crumpler, J. Grant, and L. Soderblom. 2005. "Water alteration of rocks and soils on Mars at the Spirit rover site in Gusev crater," *Nature* 436: 66–69; Bullock M.A. 2005. "Mars: the flow and ebb of water," *Nature* 438: 1087–88; McCollom T.M., and B.M. Hynek. 2005. "A volcanic environment for bedrock diagenesis at Meridiani Planum on Mars," *Nature* 438: 1129–31; Forget F., R.M. Haberle, F. Montmessin, B. Levrard, and J.W. Head. 2006. "Formation of glaciers on Mars by atmospheric precipitation at high obliquity," *Science* 311: 368–71.

69. Meierhenrich U.J., G.M. Muñoz Caro, J.H. Bredehöft, E. K. Jessberger, and W. H.-P. Thiemann. 2004. "Identification of diamino acids in the Murchison meteorite," *Proceedings of the National Academy of Sciences U S A* 101: 9182–86; Muñoz Caro G.M., U.J. Meierhenrich, W.A. Schutte, B. Barbier, A. Acones Segovia, H. Rosenbauer, W. H.-P. Thiemann, A. Brack, and J.M. Greenberg. 2002. "Amino acids from ultraviolet irradiation of interstellar ice analogues," *Nature* 416: 403–6; Bernstein M.P., J. P. Dworkin, S.A. Sandford, G.W. Cooper, and L.J. Allamandola. 2002. "Racemic amino acids from the ultraviolet photolysis of interstellar ice analogues," *Nature* 416: 401–3.

70. Meierhenrich U.J., L. Nahon, C. Alcaraz, J.H., Bredehöft, S.V. Hoffmann, B. Barbier, and A. Brack. 2005. "Asymmetric vacuum UV photolysis of the amino acid leucine in the solid state," *Angewandte Chemie International Edition* 44: 5630–34.

Chapter 3

1. The nature of the first molecules from living organisms, dating from 3.8 billion years ago, and of the first fossils of bacteria, dating from nearly 3.5 billion years ago, has been heatedly debated since 2002. See chapter 2, notes 10–14, which detail the initial publications as well as the challenges and responses to the arguments.

2. For the sake of convenience, I have used the name *chemobacteria* to emphasize how they function. The great majority of these bacteria are Archaea. Since their discovery in the 1970s, the number of species of identified Archaea has continued to grow and their remarkable properties are increasingly well understood.

3. These are essentially the Archaea that resist intense heat. The other bacteria (the eubacteria, with a few exceptions like the genera *Aquifex* and *Thermotoga*) and all the eukaryotes cannot survive prolonged exposure to greater than 70°C. The history of the main scientists who discovered the bacteria resistant to extreme heat and their research is described in an exciting book by Patrick Forterre (Forterre P. 2007. *Microbes de l'enfer* [Paris: Belin—Pour la Science]).

4. These hyperthermophilic bacteria have been found in profusion in drilling conducted in the sea off Seattle and Vancouver, in a zone of elevated magmatic activity. The record for heat resistance is 113°C for a strain of *Pyrolobus fumarii* growing in an Italian thermal

vent (discovered by Karl O. Stetter). Another record is held by strain "121" of Archaea that grows at 121°C (it can even survive for 2 hours at 130°C); this was found in Pacific submarine chimneys (Strait of Juan de Fuca, North America). This last record, published by Derek Lovley and Kazem Kashefi, of the University of Massachusetts, is debated. (Stetter, K.O. 1999. "Smallest cell sizes within hyperthermophilic Archaea ['archaebacteria']," in Space Studies Board. Commission on Physical Sciences, Mathematics, and Applications. *Size limits of very small microorganisms, Proceedings of a Workshop*. Washington, DC: National Academy Press, 68–73; and Kashefi, K., and D.R. Lovley. 2003. "Extending the upper temperature limit for life," *Science* 301: 934). Another record even more astounding is claimed by bacteria collected at 340°C and maintained in culture at 250°C under a pressure of 265 atmospheres. This record has been challenged. However, the prestigious journal *Nature*, which has its manuscripts refereed by experts in the field, published this observation. This can be seen as proof that the archaebacteria are so astounding that scientists are allowed to attribute implausible properties to them (Baross, J.W., and J.W. Deming. 1983. "Growth of black smoker bacteria at temperatures of at least 250°C," *Nature* 303: 423; Trent, J.D., R.A. Chastain, and A.A. Yayanos. 1984. "Possible artefactual basis for apparent bacterial growth at 250°C," *Nature* 307: 737–40; and Baross, J.A., and J.W. Deming. 1984. "Reply to Trent et al.," *Nature* 307: 740).

5. These archaebacteria are known as *psychrophiles*. In 1998, a research team described the oasis of bacterial life under the ice in one of the coldest regions on Earth, the Antarctic (Priscu J.C., C.H. Fritsen, C.E. Adams, S.J. Giovannoni, H.W. Paerl, C.P. McKay, P.T. Doran, D.A. Gordon, B.D. Lanoil, and J.L. Pinckney. 1998. "Perennial Antarctic lake ice: an oasis for life in a polar desert," *Science* 280: 2095–98).

6. These archaebacteria are known as *acidophiles*.

7. These archaebacteria are known as extreme *halophiles*.

8. These archaebacteria are known as *methanogens*.

9. The American microbiologist Carl R. Woese discovered this peculiarity in 1970. They were called *extremophiles*, and he observed that they had traits very different from other bacteria. They were baptized *archaebacteria,* or *Archaea*. These chemobacteria of extreme environments have heat-resistant walls, ribosomes, DNA, and a genetic code distinct from that of all other bacteria, which are known as *eubacteria* (Woese C.R. 1987. "Bacterial evolution," *Microbiological Reviews* 51: 221–71).

10. Stevens T.O., and J.P. McKinley. 1995. "Lithoautotrophic microbial ecosystem in deep basalt aquifers," *Science* 270: 450–54.

11. Gold T. 1992. "The deep, hot biosphere," *Proceedings of the National Academy of Sciences U S A* 89: 6045–49.

12. Gold, T. 1999. *The Deep Hot Biosphere: The Myth of Fossil Fuels* (New York: Springer-Verlag [Copernicus Books]).

13. This corresponds to the Archaean era.

14. So-called plant-like bacteria include all the photosynthetic bacteria that are part of the eubacteria. In this group are found rare species among the purple bacteria whose archaic chlorophyll (*bacteriochlorophyll*) is located in vesicles, and the chlorobacteria, as well as many cyanobacteria. This last group is composed of nearly 2,000 species whose chlorophyll is located on internal membranes (*thylakoids*). In these species, the green color of chlorophyll is masked by other pigments, the phycobiliproteins, yielding blue-green or purple colors.

15. I have called the ensemble of bacteria (eubacteria) that nourish themselves with organic matter *fungal bacteria*. These heterotrophs are primary decomposers, parasites, or symbionts.

16. Actinomycetes are the most common filamentous, fungus-like bacteria.

17. Modern bacterial nomenclature was born January 1, 1980, when the approved lists of bacterial names were published in the *International Journal of Systematic Bacteriology*. An International Committee on Systematics of Prokaryotes determines the validity of new taxa by applying the international code of nomenclature for prokaryotes. A major effort by bacteriologists led to a decline in the number of distinct named taxa from 29,000 to 2,336. However, the number now grows by 300 to 500 species per year. A Web site allows interested parties to follow the dynamic evolution of validated biodiversity of bacteria. Known as LBSN (List of Bacterial names with Standing in Nomenclature), it is regularly updated by Jean-Paul Euzéby (http://www.bacterio.cict.fr). Regularly updated manuals give specialists the latest fine points of bacterial taxonomy. See the following works: Garrity G.M., ed. 2001–2003. *Bergey's Manual of Systematic Bacteriology*, 2nd ed., 5 vols. (New York: Springer-Verlag); Holt J.G., ed. 2002. *Bergey's Manual of Determinative Bacteriology*, 9th ed. (New York: Springer-Verlag).

18. A total of 4,000 very different genomes are present in a 1-gram soil sample (Torsvik V., J. Goksoyr, and F.L. Daae. 1990. "High diversity in DNA of soil bacteria," *Applied and Environmental Microbiology* 56: 782–87).

19. Whitman W., D. Coleman, and W. Wiebe. 1998. "Prokaryotes: the unseen majority," *Proceedings of the National Academy of Sciences U S A* 95: 6578–83; McInerney J.O., M. Mullarkey, M.E. Wernecke, and R. Powell. 2002. "Bacteria and Archaea: molecular techniques reveal astonishing diversity," *Biodiversity* 3: 3–10.

20. The most common are *Rivularia*.

21. The most common are *Lyngbya*.

22. Several species of *Oscillatoria* are endowed with synchronous contractions that confer mobility on the cellular filament.

23. Schopf, J.W., "When did life begin?" in *Life's Origin*.

24. Viruses, plasmids, or cytoplasmic bridges (pilli) between two bacteria are the vectors for gene transfers.

25. The characters that seem to be acquired in bacteria under strong environmental selection are almost certainly present because of rapid mutations followed by transfers of modified genes. The very elevated number of individuals and the very short generation length can strengthen the impression of acquisition of new characters in a few hours or days.

26. Woese C.R. 1981. "Archaebacteria," *Scientific American* 244: 94–106.

27. An ancestral relationship of these groups with the prokaryotes was proposed by Carl R. Woese, who was the first to distinguish the archaeabacteria and the eubacteria (Woese C.R. 1993. "Universal phylogenetic tree in rooted form," *Microbiological Reviews* 58: 1–9).

28. Stephen J. Gould defended this opinion in a remarkable way. See Gould S.J. 1991. *Life's Grandeur; The Spread of Excellence from Plato to Darwin* (London: Hutchinson Radius; U.S. edition, *Full House: The Spread of Excellence from Plato to Darwin* [New York: Harmony Books]; see especially chapter 14).

29. The number of bacterial species has been estimated to be between 1 million and 100 million. See Hammond P.M. 1995. "Described and estimated species numbers: an objective assessment of current knowledge," in D. Allsop, D.L. Hawksworth, and R.R. Colwell, eds., *Microbial Diversity and Ecosystem Function* (Wallingford, UK: CAB International), 29–71.

30. This is tallied in the billions of billions of billions of individuals, so the number of individual bacteria on Earth has been estimated as 6×10^{30}, that is, the astronomic number

of 6 thousand billion billion billion individuals. If we take a gram of soil in the palm of our hand, we have gathered about 10 billion bacteria (see Whitman et al., "Prokaryotes: the unseen majority").

31. Vreeland R.H., W.D. Rozenzweig, and D.W. Powers. 2000. "Isolation of a 250 million year-old halotolerant bacterium from a primary salt crystal," *Nature* 407: 897–900. These recent discoveries of "resuscitated" bacteria have been contested, but some of the original authors remain convinced of their position (see chapter 2, note 61).

Chapter 4

1. In fact, this is only a reference to the very popular but inaccurate notion of realms of life. Though the fungi were considered until the late twentieth century to be plants, genetic research now shows that fungi are closer to animals.

2. This is the case especially among the ascidians and urochordates or tunicates, invertebrates that preceded vertebrates.

3. Meinesz A. 1999. *Killer Algae: The True Tale of a Biological Invasion* (Chicago: University of Chicago Press; French edition: 1997. *Le roman noir de l'algue "tueuse."* Caulerpa taxifolia *contre la Méditerranée* [Paris: Belin]).

4. In Australia, 24 million hectares of *Opuntia* were almost completely destroyed by the caterpillars of the small South American moth *Cactoblastis cactorum*.

5. Several publications discuss dangers of biological control. Center T.G., J.H. Frank, and F.A. Dray Jr. 1997. "Biological control," in D. Simberloff, D.C. Schmitz, and T.C. Brown, eds., *Strangers In Paradise: Impact and Management of Nonindigenous Species in Florida* (Washington, DC: Island Press), 245–63; Simberloff D., and P. Stiling. 1996. "How risky is biological control?" *Ecology* 77: 1965–74; Simberloff D., and P. Stiling. 1996. "Risks of species introduced for biological control," *Biological Conservation* 78: 185–92; Stiling P. 2002. "Non-target effects of the biological control agent *Cactoblastis cactorum* in the U.S. and beyond," *Biological Invasions* 4: 273–81.

6. Thibaut T., A. Meinesz, P. Amade, S. Charrier, K, De Angelis, S. Ierardi, L. Mangialajo, J. Melnick, and V. Vidal. 2001. "*Elysia subornata* (Mollusca) a potential control agent of the alga *Caulerpa taxifolia* (Chlorophyta) in the Mediterranean Sea," *Journal of the Marine Biological Association UK* 81: 497–504.

7. Stoecker D.K., A.E. Michaels, and L.H. Davis. 1987. "Large proportion of marine planktonic ciliates found to contain functional chloroplasts," *Nature* 326: 790–92.

8. Instead of *symbiosis,* parasitologists prefer the term *mutualism.* A very detailed book by the French parasitologist Claude Combes discusses and synthesizes the various semantic approaches to all kinds of associations among living organisms (Combes C. 2001. *Parasitism: The Ecology and Evolution of Intimate Interactions* [Chicago: University of Chicago Press]).

9. Among the eukaryotes, only rare parasites can live in an environment without oxygen (that is, anaerobically).

10. See chapter 2, notes 31 and 32 for a more detailed explanation.

11. Among the pioneers of the theory of endosymbiosis are several Russian botanists (C. Merezhkovsky, I.E. Wallin, A.S. Famintsyn, and B.M. Kozo-Polyansky). Several works describe the history of their approach. See Khakhina L.N. 1979. *Concepts of Symbiogenesis* (in Russian) (Leningrad: Akademie Nauk, USSR [Soviet Academy of Sciences]; English language edition: 1992. *Concepts of Symbiogenesis: A Historical and Critical Study of the Research of Russian Botanists.* L. Margulis and M. McMenamim, eds. [New Haven, CT: Yale University Press]). All these authors conceived of a process of endosymbiosis with very few

elements (Schimper A.F.W. 1883. "Über die entwicklung der chlorophyll körner und farbkörner," *Botanische Zeitung* 41: 105–14; Merezhkovsky C. 1905. "Über die Natur und den Ursprung der Chromatophoren im Pflanzenreiche," *Biologisches Centralblatt* 25: 593–604; Merezhkovsky C. 1910. "Theoric dcr zwei plasmaarten als grundlage der symbiogenesis, einer neuen lehre von der entstehung der organismen," *Biologisches Centralblatt* 30: 278–303, 321–47, 353–67). See also Wallin I.E. 1927. *Symbionticism and the Origin of Species* (Baltimore: Williams and Wilkins).

12. Also called SET (serial endosymbiosis theory).

13. Lynn Margulis has published many works, the last with her elder son Dorion Sagan. In one of these books, she describes the adversity she faced for having published her theory, including rejected manuscripts and heavily criticized publications (Margulis L. 1998. *Symbiotic Planet: A New Look at Evolution* [New York: Basic Books]). Margulis's first publication (published under L. Sagan, her married name) on this subject dates from 1967 (Sagan L. 1967. "On the origin of mitosing cells," *Journal of Theoretical Biology* 14: 255–74). Her first book sketched the main lines of the theory (Margulis L. 1970. *Origin of Eukaryotic Cells: Evidence and Research Implications for a Theory of the Origin and Evolution of Microbial, Plant, and Animal Cells on the Precambrian Earth* [New Haven, CT: Yale University Press]). Her first popular publication was "Symbiosis and Evolution," *Scientific American* 225 (1971): 48–57. Other major works of hers include: *Symbiosis in Cell Evolution: Life and its Environment on the Early Earth* (San Francisco: W.H. Freeman, 1981); *Symbiosis in Cell Evolution: Microbial Communities in the Archean and Proterozoic Eons* (New York: W.H. Freeman, 1993); and Margulis L., and D. Sagan. 2002. *Acquiring Genomes: A Theory of the Origin of Species* (New York: Perseus).

14. Mycoplasmas, bacteria that are now parasites, are a good example of bacteria deprived of a rigid wall.

15. This is the cytoskeleton, which is common to all eukaryotes. It is a network of intracellular fibers made of three large families of filamentous proteins. They are thick (filaments of tubulin, or microtubules), thin (constructed of actin or micro-filaments), or have an intermediate structure.

16. It is mainly the myxomycetes that behave this way.

17. This is phagocytosis.

18. Mitochondria derive from present-day alpha protobacteria (which include *Rickettsia prowazekii,* the pathogen that causes typhus), which enter parasitized host cells during phagocytosis.

19. Martin W., and M. Müller. 1998. "The hydrogen hypothesis for the first eukaryote," *Nature* 392: 37–41.

20. The precursors of mitochondria are the hydrogenosomes.

21. Animal protists (protozoans).

22. Mitochondria are absent from certain unicellular parasites that are sheltered from oxygen. But in the genomes of these diversified cells, parasites of multicellular organisms that appeared less than 600 million years ago, traces remain of the co-opting of the bacteria that became mitochondria.

23. In the glaucophytes, a phylum of unicellular algae, the chloroplast is still surrounded by a wall of the bacterial type.

24. This is observed in the cryptophytes. In representatives of this phylum of unicellular algae, the subordinate, atrophied nucleus is termed a nucleomorph.

25. These are bacteriophages and conjugative plasmids.

26. This is mitosis.

27. This is meiosis, also called reduction division or chromosome reduction. In a complete life cycle of an animal or plant, at a particular moment, there is a marriage of similar (homologous) chromosomes (meaning fertilization leading to a doubling of chromosome number), which is followed by the diploid phase of the life cycle. At another programmed moment, there is a divorce (which leads through meiosis to the production of gametes each having a single set of chromosomes), followed by the haploid phase of the life cycle.

28. The suffix *gamy* (as in monogamy or bigamy) means "marriage."

29. Maynard Smith J., and E. Szathmary. 1999. *The Origin of Life: From the Birth of Life to the Origin of Language* (Oxford: Oxford University Press).

30. Lynn Margulis and Dorian Sagan propounded the theory of the appearance of sexual reproduction in cannibalistic protists in their book *Origins of Sex: Three Billion Years of Genetic Recombination* (New Haven: Yale University Press, 1986).

31. This fusion of nuclei is called *karyogamy,* which means "marriage of nuclei."

32. A large majority of edible mushrooms are basidiomycetes.

33. The phenomenon of cellular fusion without fusion of the nuclei is called *somatogamy.*

34. The secondary mycelia are formed from cells called *dikaryons* (dikaryotic cells); they contain two nuclei.

35. Exchanges of segments of homologous chromosomes occur during the prophase of the first meiotic division. In 1910, the Nobel laureate Thomas Hunt Morgan (1866–1945) elaborated the concept of *crossing over* to describe the recombination of chromosomes.

36. There is a different theory about the acquisition of flagella that stipulates that the origin of eukaryotes came about by the symbiotic association between an archaebacterium similar to modern-day *Thermoplasma* and a eubacterium similar to the modern-day *Spirochaetes*. According to the theory, this chimeric ensemble would have constituted an amoeboid prototype of a eukaryote, without mitochondria but with two flagella and a nucleus. This archaic amoeba was then a partner in subsequent endosymbioses that integrated the mitochondrion and the chloroplast (Margulis L., M.F. Dolan, and R. Guerrero. 2000. "The chimeric eukaryote: origin of the nucleus from the karyomastigont in amitochondriate protists," in F.J. Ayala, W.M. Fitch, and M.T. Clegg, eds., *Variation and Evolution in Plants and Microorganisms: Toward a New Synthesis 50 Years after Stebbins* [Washington, DC: National Academy Press], 21–34).

37. The spirochetes, minuscule helical bacteria, are mobile thanks to their cytoskeleton. Among the spirochetes, one species is particularly well known: *Treponema pallidum,* the pathogen that causes syphilis.

38. The phyla whose representatives show no trace of flagella are the red algae (Rhodophyta) and the ensemble of "septomycete" fungi (ascomycetes and basidiomycetes), also classified in the phylum eumycetes.

39. Two examples of unclassifiable chimerae, dating from 2.1 billion years ago, are the umbrella-shaped *Kakabekia umbellata* and *Eosphaera,* a tiny globe surrounded by still tinier globes (possibly reproductive bodies), the whole of which is encased in a mucilaginous mass. Photographs of these strange organisms (chapter 6, figs. 15a and 16a) can be found in *Cradle of Life,* by J. William Schopf.

40. The American paleontologist David M. Raup has published calculations of the survival rates of various taxonomic levels of biodiversity (phyla, classes, orders, families, species, and so on) faced with a cataclysm. In his view, life has passed through several apocalyptic episodes, analogous to a platoon passing through a field of bullets. He evaluates

the chances of survival of the various levels of biodiversity according to the "density of bullets" during the five cataclysms that have devastated life since the rapid emergence of animals in the Cambrian. At the time of emergence of the eukaryotes by endosymbiosis, and during the two billion years of evolution that followed, membership in the various levels must have dwindled. Among the many different endosymbiotic associations, all living in a relatively homogeneous habitat, essentially by floating just beneath the surface of the sea, each could potentially have created a distinct phylum. We can deduce that the Precambrian cataclysms had a much greater impact than the impacts of the last five great cataclysms on the essentially random selection of surviving phyla (each with its organizational structure). Otherwise, each phylum would already have had multiple forms adapted to different habitats, which would have favored the survival of at least some representatives of each phylum (Raup D.M. 1991. *Extinction: Bad Genes or Bad Luck?* [New York: W. W. Norton]).

41. These decorative brown algae (Phaeophyta) most often belong to the genera *Fucus* and *Ascophyllum*.

42. The phyla Dinophyta and Euglenophyta exemplify this situation. Each has representatives that are autotrophs and others that are heterotrophs.

43. In fact, this is a case of chloroplast theft: "kleptoplastia"! By coevolution, the sea slugs have succeeded in taking for their own benefit the chloroplasts, ancient cyanobacteria "enslaved" by green plants since the genesis of the latter (the first eukaryotes arose by endosymbiosis 2.7 billion years ago; see chapter 6, note 10). Well tamed by the green plants, they were then stolen by the sea slugs, who contrived to make them function in their bodies without allowing them to reproduce.

44. Coevolution (a new word coined by biologists) entails a succession of reciprocal selective pressures exerted by two species on one another through their interaction. In the course of evolution, each constantly acquires new defensive or offensive weapons in order not to be overtaken by the other in an endless arms race.

Chapter 5

1. Only three Vermeer paintings are signed and dated.

2. The instruments represented are the guitar, cither, lute, trumpet, bass viol or viola de gamba (types of contrabasses), virginal (a type of spinet or harpsichord), and harpsichord.

3. Five different maps served as models.

4. I have visited old Delft. Van Leeuwenhoek's house no longer stands and Vermeer's has been transformed into two antique stores. I noted that the two houses were located very close to each other and that Delft in 1650 was only a small village that one could traverse on foot in less than half an hour. Knowing that the painter and the scientific autodidact were the same age and passed their entire lives in Delft, I understood how they must have known one another and how Vermeer must have been interested in Van Leeuwenhoek's research.

5. Three recent books are dedicated to the life and work of Van Leeuwenhoek, and a collection of all his letters has also been published. Ford B.J. 1991. *The Leeuwenhoek Legacy* (Bristol: Biopress; London: Farrand) (this author tells in particular of his discovery of preparations for microscopes in Van Leeuwenhoek's letters kept in the archives of the Royal Society of London); Boutibonnes P. 1994. *Van Leeuwenhoek, l'excellence du regard* (Paris: Belin) (this book ably documents Van Leeuwenhoek's research and its context); Yount L. 1996. *Antoni Van Leeuwenhoek: First to See Microscopic Life*, Great Minds of Science (Berkeley Heights, NJ: Enslow).

6. *Monad* signifies simple and unable to be broken down into components. Plato first used this term to mean *idea*. Gottfried Leibniz (1645–1716) was the first to define a monad as a simple substance in his book, *Monadology* (1714).

7. We attribute to Theodor Schwann (1810–1882), German physiologist, one of the founders of modern histology, the observation that all animals are composed of cells, and to the German botanist Matthias Schleiden (1804–1881) the observation that all plants are composed of cells.

8. Stephen J. Gould consecrated an entire chapter to the explication of the term *evolution* and the history of its use in his book *Ever Since Darwin: Reflections in Natural History* (New York: W. W. Norton, 1977 [see especially chapter 3]).

9. A pertinent analysis of the "war of ideas" that shook American intellectual circles with respect to evolutionary theories and their philosophical and social implications was published by a philosopher who specializes in the subjective penchants of the main evolutionary theorists. See Ruse M. 1999. *Mystery of Mysteries: Is Evolution a Social Construction?* (Cambridge, MA: Harvard University Press).

10. The conceptual stages that improved Darwin's initial theory are designated as the neodarwinian synthesis, which incorporated population genetics into evolution, and the modern synthesis or the synthetic theory of evolution, which corresponds to the progressive integration of various biological disciplines into an evolutionary schema (caryology, embryology, molecular genetics, and genomics). Four authors in particular contributed to the synthetic theory of evolution: Theodosius Dobzhansky (1900–1975), Ernst Mayr (1904–2005), George Gaylord Simpson (1902–1984), and G. Ledyard Stebbins (1906–2000). They published their main concepts between 1937 and 1950 in magisterial books by the same publisher. Dobzhansky T. 1937. *Genetics and the Origin of Species* (New York: Columbia University Press); Mayr E. 1942. *Systematics and the Origin of Species: A Viewpoint of a Zoologist* (New York: Columbia University Press); Simpson G.G. 1944. *Tempo and Mode in Evolution* (New York: Columbia University Press); Stebbins G.L. 1950. *Variation and Evolution in Plants* (New York: Columbia University Press).

11. The philosopher Jean Gayon published a small relevant article on these terms, which are too often used incorrectly in the evolution literature (Gayon J. 2001. "Hasard et évolution," in H. Le Guyader, ed., *L'évolution* [Paris: Belin—Pour La Science], 12–14). Stephen J. Gould has also written on their often ambiguous and counter-intuitive meanings (Gould, *Eight Little Piggies* [see especially chapter 29]).

12. The French word *hasard* (good or bad luck) comes from the Arab *az-zahr,* which means game of dice. This is also the etymological source of the English word *hazard.*

13. To describe this luck, the Nobel laureate Jacques Monod uses the definition of *absolute coincidence*. Monod J. 1970. *Le hasard et la nécessité. Essai sur la philosophie naturelle de la biologie moderne* (Paris: Editions de Seuil; English language edition: 1972. *Chance and Necessity: An Essay on the Natural Philosophy of Modern Biology* [New York: Vintage Books]).

14. The French word *aléa* (chance, hazard, or risk), connoting an unforeseeable event, derives from the Latin *alea,* a game of dice (*aleatorius* and *aleator* refer to players of dice; see also the English word *aleatory*).

15. *Contingent* comes from the Latin *contingens; contingere* means happening by chance. The principle of contingency, used to describe the succession of random events that marks the evolution of living organisms, was mentioned first by the Nobel laureates Jacques Monod and François Jacob in their respective works, both appearing in 1970 (Jacob F. 1970. *La logique du vivant* [Paris: Gallimard; English language edition: 1982. *The Logic of Life:*

A History of Heredity, New York: Pantheon Books]; Monod, *Le hasard et la nécessité*). In 1989, the paleontologist Stephen J. Gould cited contingent events as the global driving principle of evolution (Gould, *Wonderful Life*).

16. Authors enjoy describing such imaginary cascades. Stephen J. Gould illustrates his ideas by a cascade beginning with the thawing of an estuary and ending in a meeting between a bear and a seal (Gould, *Eight Little Piggies* [see especially chapter 29]). André Langaney imagined another little story of cascades of unlikely events ending with the death of "Madame Durand" (Langaney, *La philosophie . . . biologique* [see p. 98]).

17. Even today, we must be overwhelmed by the simplicity of the presentation of how the structure of DNA was discovered: it is shown on a single page illustrated by a simple figure. Watson J.D., and F.H.C. Crick. 1953. "Molecular structure of nucleic acids: a structure for desoxyribose nucleic acid," *Nature* 4356: 737.

18. The term *reductionism* appears in biology books in the late 1960s. See Koestler A., and J.R. Smithies, eds. 1969. *Beyond Reductionism: New Perspectives in the Life Sciences* (London: Hutchinson). In France, the term *reductionism* was first used by the Nobel laureates François Jacob and Jacques Monod (Jacob, *La logique du vivant*; Monod, *Le hasard et la nécessité*). The concept was subsequently defended by the president of the Academy of Sciences of Paris, François Gros, in his book *Les secrets du gène* (Paris: Odile Jacob-Seuil, 1986).

19. René Descartes, a contemporary of Vermeer and Van Leeuwenhoek, enlisted in the army of the princes of Nassau to battle the Spanish invaders in the United Provinces of the Netherlands. He wrote his most beautiful texts during his long sojourn in Holland, where he found peace far from persecutors. His *Discours de la méthode* was initially published in Leiden, then a small Dutch town less than 20 kilometers from Delft (Descartes R. 1637. *Discours de la méthode pour bien conduire sa raison et chercher la vérité dans les sciences* [Leiden: Imprimerie Ian Maire; English language edition: 1956. *Discourse on Method*, New York: Liberal Arts Press]).

20. A Dutch botanist, Hugo de Vries (1848–1935), first discovered rapid mutations in plants. Between 1901 and 1903 he published a book on the subject, *The Theory of Mutation* (2 vols. Leipzig: Veit & Co.; see also H. de Vries. 1909–1910. *The Mutation Theory: Experiments and Observations on the Origin of Species in the Vegetable Kingdom,* 2 vols. [Chicago: Open Court Publishing Company]). He was followed by Thomas Hunt Morgan (1866–1945), who worked on fruit flies (*Drosophila*). For his research, which contributed heavily to the chromosomal theory of heredity, Morgan received the Nobel Prize in Physiology and Medicine in 1933 (Morgan T.H. 1910. "Sex-limited inheritance in *Drosophila*," *Science* 32: 120–22; idem. 1911. "The origin of nine wing mutations in *Drosophila*," *Science* 33: 496–99; idem. 1911. "The origin of five mutations in eye color in *Drosophila* and their mode of inheritance," *Science* 33: 534–37.

21. Monod, *Le hasard et la nécessité.*

22. Micromutations cause changes within a species.

23. Macromutations cause changes that can generate a change in species (a jump or saltation from one species to another).

24. This is the neutral theory of molecular evolution initiated by the Frenchman Gustave Malécot (1911–1998) and developed by the Japanese geneticist Motoo Kimura (1924–1994). Kimura M. 1968. "Evolutionary rate at the molecular level," *Nature* 217: 624–26; idem. 1983. *The Neutral Theory of Molecular Evolution* (Cambridge: Cambridge University Press); idem. 1991. "The neutral theory of molecular evolution: a review of recent evidence," *Japanese Journal of Genetics* 66: 367–86; idem. 1998. "La théorie neutraliste de

l'évolution moléculaire" (posthumous publication), in H. Le Guyader, ed., *L'évolution*, 150–59.

25. Mendel discovered the laws of heredity. But the significance of his studies, published in 1866, was not recognized until the beginning of the twentieth century, in publications by Carl Correns (1864–1933), Erich von Tschermak (1871–1962), and Hugo de Vries (1848–1935). See Le Guyader H., and J. Génermont. 2001. "L'évolution: une histoire des idées," in Le Guyader, ed., *L'évolution*, 12–14.

26. A German biologist, August Weismann (1834–1914), first explained the laws of heredity by the association of chromosomes and exchanges of portions of them, just before the formation of eggs and sperm.

27. In 1908 and 1909, an English mathematician, Harold Hardy (1877–1947), and a German physician, Wilhelm Weinberg (1862–1937), independently formulated an "equilibrium" law, according to which allele frequencies remain unchanged from generation to generation within an "ideal" diploid population and depend only on the frequencies in the initial generation (Hardy G.H. 1908. "Mendelian proportions in a mixed population," *Science* 28: 49–50; Weinberg W. 1908. "Über den nachweis der verebung beim menschen," *Jahreshelfe des Vereins für Vaterländische Naturkunde in Württemberg* 64: 368–82; idem. 1909. "Über vererbungsgesetze beim menschen," *Zeitschrift für Induktive Abstammungs- und Vererbungslehre* 1: 377–93 and 2: 440–60).

28. Pierre Teilhard de Chardin wrote ten books on this subject, including: *Le phénomène humain*; *L'apparition de l'Homme* (Paris: Seuil, 1956; trans. by J. M. Cohen as *The Appearance of Man* [New York: Harper & Row, 1965]); *La place de l'Homme dans la nature. Le groupe zoologique humain* (Paris: Seuil, 1965; trans. by R. Hague as *Man's Place in Nature: The Human Zoological Group* [New York: Harper & Row, 1966]).

29. Henri Bergson (1859–1941) developed the notion of the continuity of life and of evolutionary progress. "Life appears as a current that runs from seed to seed by the intermediary of the fully developed organism. Everything happens as if the organism itself was but an excrescence, a bud that puts forth the ancient seed, working to perpetuate itself as a new seed. The essential point is the indefinite continuity of progress, invisible progress in which each visible organism spans just the short interval of time given to it to live" (Bergson H. 1907. *L'évolution créatrice* [Paris: Librairies Félix Alcan et Guillaumin réunies; English language edition: 1911. *Creative Evolution*, New York: H. Holt & Co.]).

30. Stephen J. Gould analyzed in detail the changing Vatican views towards the scientific principles of evolution in his book *Rocks of Ages: Science and Religion in the Fullness of Life* (New York: Random House, 1999). With the accession of Pope Benedict XVI and the growth of the intelligent design movement, there is new opposition within the church to certain ideas on the origin and evolution of life. The stance of the church on these matters is discussed in chapter 9.

31. It took nearly two centuries to accept definitively Copernicus's theory that the earth revolves around the sun. This statement appeared the day of his death in 1543 in the book *De revolutionibus orbitum coelestium libri,* seemingly destined for Pope Paul III (1468–1549). This pope, very open to scientific ideas, accepted it. But subsequently, doubt gained the upper hand. Obscurantism returned with the torture of Giordano Bruno in 1600 (he had held that the earth moved and the universe is infinite) and Galileo's recantation, obtained by Pope Urban VIII (1568–1644). More than a century later, in 1741, Pope Benedict XIV (1675–1758) recognized heliocentrism.

32. Teilhard de Chardin called this type of evolution *orthogenesis,* and he termed its goal the *omega point.*

33. In their own way, the Mormons use a text from the Bible: as the Lord in fact teaches, the dead as well as the living should be baptized if they want to enter the kingdom of God (see John 3:1–5 and Corinthians 15:29).

34. Allah says, "Rush to the mercy of your Lord and towards a garden, as large as the heavens and earth, prepared for those who fear Allah" (verset 133/9).

35. Monod spoke of teleonomy, of teleonomic life (*teleo* comes from the Greek *teleo* or *teleos* (end, goal) and *teleios* (finished, completed). This suggests that living organisms have a goal and work towards an end (Monod, *Le hasard et la nécessité*).

36. Jacob, *La logique du vivant*.

37. Dawkins, *The Selfish Gene*.

38. Ridley M. 1996. *Evolution*. 2nd ed. (Oxford: Blackwell).

39. Crick F., and L. Orgel. 1980. "Selfish DNA: the ultimate parasite," *Nature* 284: 604–7.

40. Dawkins R. 2001. "La loi des gènes. Pourquoi la vie? Parce qu'elle assure la survie des gènes," in H. Le Guyader, ed., *L'évolution*, 18–24

41. It was while reading Malthus's "Essay on the principle of population" that Darwin had the insight that led to the essential principle. (Malthus, T.R. 1798. "An essay on the principle of population, as it affects the future improvement of society with remarks on the speculations of Mr. Godwin, M. Condorcet, and other writers" [London: Johnson]).

42. The fundamental influence of the differential rates of reproduction in evolution is one of Darwin's crucial contributions.

43. In 1917, English biologist D'Arcy Wentworth Thomson (1860–1948) published a remarkable book that made biologists understand the importance of physical and chemical constraints in which forms evolve (deducible, according to him, by the laws of mathematics and geometry). Thomson D.W. 1992. *On Growth and Form*. Abridged ed. (Cambridge: Cambridge University Press).

44. Dawkins R. 1986. *The Blind Watchmaker* (New York: W. W. Norton).

45. Stephen J. Gould notes the analogizing of natural selection to a composer (by Theodosius Dobzhansky), a poet (by George Gaylord Simpson), a sculptor (by Ernst Mayr), and even Shakespeare (by Julian Huxley). Gould, *Ever Since Darwin* (see especially chapter 4).

46. It is especially the blind (or automatic) determinism of the mechanism of cumulative natural selection that Dawkins uses to support his metaphor of the blind watchmaker, as opposed to the absolute randomness of the situation, of the conditions in which a modification appears that has been produced by new genetic information (Dawkins, *The Blind Watchmaker*).

47. Stephen J. Gould discovered that the work sent by Karl Marx to Darwin had not been read in its entirety, because many pages were still attached to one another, not having been cut by the printer (Gould, *Ever Since Darwin*; see chapter 1). In this book, Gould reported a legend to the effect that Marx wanted to dedicate his book to Darwin, who politely refused. Twenty-five years later, Gould returned to this story, explaining that a mix-up in addressees of letters had incorrectly led a biographer of Marx to publish this story. (Gould, *I Have Landed*; see chapter 6).

48. He wrote 22 of these (more than 5,000 pages total), including a huge synthesis of evolution, which alone was more than 1,400 pages, just before he died (Gould, S.J. 2002. *The Structure of Evolutionary Theory* [Cambridge, MA: Belknap Press of Harvard University Press]).

49. Analysis of citations of Gould's main works by professional scientists shows that his popular books are rarely cited and that his academic works are far less cited than those of Edward O. Wilson, celebrated author of magisterial works concerning biodiversity and animal behavior, especially sociobiology (Ruse, *Mystery of Mysteries*; see chapter 7). In fact, the urge to popularize the science of evolution seems to have been an obsession of Gould's. He justified the personal writing style of his essays in the prologues of three of these popular books: *The Lying Stones of Marrakech*, *Dinosaur in a Haystack*, and *Bully for Brontosaurus*. In this last book, he praises the French tradition of essays, begun by Montaigne. He regrets and profoundly deplores the fact that, in the United States, popular scientific books are deprecated and variously called adulterations, simplifications, deformations towards sensationalism, research into the spectacular, and pathways to commercial success.

50. The American literary critic Frederick C. Crews has written two brilliant reviews of a dozen books appearing near the end of the twentieth century in which the meaning of Darwinism is discussed in contradictory fashion. Among the authors who have contributed to enriching the debate are three Americans: Michael Ruse, an agnostic philosopher, John F. Haught, a theologian, and Kenneth R. Miller, a Darwinian biologist who is nevertheless a religious believer. They have recently presented a diversity of strong arguments to prove that Darwinism is compatible with Christianity (Crews F.C. 2001. "Saving us from Darwin," *New York Review of Books* 48: 15[I] and 16[II]; Ruse M. 2001. *Can a Darwinian Be a Christian? The Relationship between Science and Religion* [Cambridge: Cambridge University Press]; Haught J.F. 2000. *God after Darwin: A Theology of Evolution* [Boulder, CO: Westview]; Miller K.R. 1999. *Finding Darwin's God: A Scientist's Search for Common Ground Between God and Evolution* [New York: Cliff Street Books/HarperCollins]).

51. Dawkins R. 2006. *The God Delusion* (Boston: Houghton Mifflin).

52. Gould, *Rocks of Ages*. Gould began by developing this idea in his book, *Leonardo's Mountain of Clams and the Diet of Worms* (New York: Harmony Books, 1998). In chapter 14, "Non-overlapping magisteria," he introduced NOMA.

53. The author of this statement is Pope John Paul II ("Magisterium Is Concerned with Question of Evolution For It Involves Conception of Man," Message of Pope John Paul II to the Pontifical Academy of Sciences meeting in plenary session. The Vatican, October 22, 1996).

54. Einstein doubted the contingencies of life; he could never admit that God plays dice. He believed the only laws of value were those in a universe "where something exists objectively" (Einstein A., M. Born, and H. Born, *Correspondences 1916–1955* [Paris: Seuil, 1972]; see the letter from Einstein dated September 7, 1944 and the reply of the wife of the physicist Max Born, October 9, 1944; American edition: 1971. *The Born-Einstein Letters: Correspondence Between Albert Einstein and Max and Hedwig Born from 1916 to 1955* [New York: Walker & Co.]).

Chapter 6

1. This is the phylogenetic classification. It is been polished during the last few decades by the introduction of molecular methods and further improved more recently by application of a method that accounts for all indices of relationship in order to create genealogical trees of life: cladistics. Brooks D.R., and D.A. McLennan. 1991. *Phylogeny, Ecology, and Behavior* (Chicago: University of Chicago Press); Harvey P.H., and M.D. Pagel. 1991. *The Comparative Method in Evolutionary Biology* (Oxford: Oxford University Press); Lecointre G., and H. Le Guyader. 2007. *The Tree of Life: A Phylogenetic Classification* (Cambridge, MA: Harvard University Press).

2. This is the Proterozoic era. Andrew H. Knoll, a leading paleontologist of the Precambrian, has divided it into 17 stratigraphic intervals (Knoll, A.H. 1995. "Proterozoic and early Cambrian protists: evidence for accelerating evolutionary tempo," in W.M. Fitch and F.J. Ayala, eds., *Tempo and Mode in Evolution* [Washington, DC: National Academy Press], 63–83.). J. William Schopf has written a brief history of the discovery of Precambrian species (Schopf J.W. 2000. "Solution to Darwin's dilemma: discovery of the missing Precambrian record of life," in Ayala, Fitch, and Clegg, eds., *Variation and Evolution in Plants and Microorganisms*, 6–20).

3. Antoni van Leeuwenhoek plays a role in a historical novel about the life of the artist (Tracy Chevalier, *Girl with a Pearl Earring* [London: Harper Collins, 1999]; a blockbuster movie of the same name was adapted from this novel by Peter Webber in 2003).

4. In one of his first works, Stephen J. Gould aptly recalled the state of mind reigning among scientists at the time when the oldest known fossils all dated back only to the Cambrian (Gould, *Ever Since Darwin* [see especially chapters 14 and 15]).

5. In 1965, Elso S. Barghoorn and Stanley A. Tyler published their discovery of fossils older than two billion years (Barghoorn E.S., and S.A. Tyler. 1965. "Microorganisms from the Gunflint chert," *Science* 147: 563–77). The saga of the discovery of the oldest fossils is described by J. William Schopf (Schopf, *Cradle of Life*). See also Knoll, *Life on a Young Planet*.

6. A wide-ranging debate on this subject was set off by contradictory publications proposing dates differing by a factor of more than 2 (see the debate on dating the first multicellular animals: notes 27 and 28 of this chapter). For dating of plants that appeared much more recently than the Cambrian, the same errors were revealed: Sanderson M.J., and J.A. Doyle. 2001. "Sources of error and confidence intervals in estimating the age of angiosperms from rbcL and 18S rDNA data," *American Journal of Botany* 88: 1499–1516.

7. Orlando L. 2005. *L'anti-Jurassic Park. Faire parler l'ADN fossile* (Paris: Belin. Pour la Science).

8. I must note here the exceptional discoveries of leaves of *Magnolia* and other plants sheltered from oxygen in mud and thus preserved for 20 million years (in lake beds near the town of Clarkia, Idaho), from which has been extracted and analyzed part of the most common gene in plants (contained in chloroplasts). Stephen J. Gould emphasizes the fact that, if these data are substantiated, they provide evidence on how long evolution has been occurring that derives from a method other than anatomical analysis of fossils. In the United States, where many creationists consider evolutionary theories to be without scientific foundation, any evidence confirming predictions of evolutionary theory is valuable (Gould S.J. 1995. *Dinosaur in a Haystack: Reflections in Natural History* [New York: Harmony Books]; see chapter 31). Although the age of the DNA from the Clarkia *Magnolia* fossils has been questioned, the most recent publication confirms the extraordinarily good preservation of DNA in these Miocene plants. See Golenberg E.M., D.E. Giannasi, M.T. Clegg, C.J. Smiley, M. Durbin, D. Henderson, and G. Zurawski. 1990. "Chloroplast DNA sequence from a Miocene *Magnolia* species," *Nature* 344: 656–58; Pääbo S., and C. Wilson. 1991. "Miocene DNA sequence—a dream come true?" *Current Biology* 1:45–46; Kim S., D.E. Soltis, P.S. Soltis, and Y. Suh. 2004. "DNA sequences from Miocene fossils: an *ndhF* sequence of *Magnolia latahensis* (Magnioliaceae) and an *rbcL* sequence of *Persea pseudocarolinensis* (Lauraceae)," *American Journal of Botany* 91: 615–20.

9. For specialists, two terms with same prefix, *proto* (from the Greek *protos,* meaning first, rudimentary, primitive), designate this simple state: protist (or *Protoctista*) designates all animals and plants composed of a single cell, and protozoan designates independent, unicellular animals.

10. In Australian schists dating from nearly 2.7 billion years ago, hydrocarbons have been found that are characteristic of biosynthesis in eukaryotes (steranes like cholestane). See Brocks J.J., G.A. Logan, R. Buick, and R.E. Summons. 1999. "Archean molecular fossils and the early rise of eukaryotes," *Science* 285: 1033–36.

11. The dating of oxidation of minerals allows the estimation on the one hand of the presence of a large number of plant-like bacteria necessary for the accumulation of oxygen in water and the atmosphere, and on the other hand the timing of the rise in oxygen concentration to the point that it allowed the emergence of unicellular and multicellular plants and animals. The very marked increase in deposits of oxidized minerals dates from 2.5 billion years ago, which suggests it is a consequence of a global spread of many types of plant-like bacteria at this time (see chapter 2, note 22). Other references are Runnegar B. 1991. "Precambrian oxygen levels estimated from the biochemistry and physiology of early eukaryotes," *Palaeogeography, Palaeoclimatology, Palaeoecology* 71: 97–111; Shen Y., R. Buick, and D.E. Canfield. 2001. "Isotopic evidence for microbial sulphate reduction in the early Archaean era," *Nature* 410: 77–81.

12. These carbonaceous fossil organisms have been named *Grypania spiralis* (Han T.M., and B. Runnegar. 1992. "Megascopic eukaryotic algae from the 2.1-billion-year-old Negaunee iron-formation, Michigan," *Science* 257: 232–35).

13. In addition to the astounding and rare unicellular forms dating from 2.1 billion years ago and later (such as the parasol cell *Kakabekia umbellata* or the cell with satellites *Eosphaera*), the most frequent forms, which appear beginning 2.15 billion years ago, are called acritarchs (fig. 15 in chapter 6). These acritarchs are almost certainly cysts (resistant resting cells) of dinophytes (a phylum of soft unicellular algae, also known as dinoflagellates, well represented today in oceans and fresh water). See Goodman D.K. 1987. "Dinoflagellate cysts in ancient and modern sediments," in F.J.R. Taylor, ed., *The Biology of Dinoflagellates: Botanical Monographs No. 21* (Oxford: Blackwell), 649–722; and Zhang W.-L. 2007. "Deposition and deformation of late Archaean sediments and preservation of microfossils in the Harris Greenstone Domain, Gawler Craton, South Australia," *Precambrian Research* 156: 107–24. The first acritarch described (in 1899), *Chuaria circularis,* was at first confounded with an animal species (a brachiopod). In 1928, it was classified as an acritarch, and it is now considered a multicellular alga. See Zhu S., X. Huang, Y. He, G. Zhu, L. Sun, and K. Zhang. 2000. "Discovery of carbonaceous compressions and their multicellular tissues from the Changzhougou formations (1800 Ma) in the Yanshan range, North China," *Chinese Science Bulletin* 45: 841–47.

14. An analysis of the acritarch flora was published by Andrew H. Knoll (Knoll, "Proterozoic and early Cambrian protists: evidence for accelerating evolutionary tempo," in Fitch and Ayala, eds., *Tempo and Mode in Evolution* [see note 2 in this chapter]) and John Warren Huntley et al. (Huntley J.W., S. Xiao, and M. Kowalewski. 2006. "1.3 Billion years of acritarch history: an empirical morphospace approach," *Precambrian Research* 144: 52–68).

15. The first fossil of a multicellular, eukaryotic species belonging to the present-day genus *Bangia,* still present today, is a red alga formed from simple filaments of cells (Butterfield N.J., A.H. Knoll, and K. Swett. 1990. "A bangiophyte red alga from the Proterozoic of arctic Canada," *Science* 250: 104–7; Butterfield N.J. 2000. "*Bangiomorpha pubescens* n. gen., n. sp.: implications for the evolution of sex, multicellularity and the Mesoproterozoic/Neoproterozoic radiation of Eucaryotes," *Paleobiology* 26: 386–404). Nicholas J. Butterfield also described the first species that are probably higher fungi (idem. 2005. "Probable Proterozoic fungi," *Paleobiology* 31: 165–82).

16. Bloeser B. 1985. "*Melanocyrillium,* a new genus of structurally complex late Proterozoic microfossils from the Kwagunt formation (Chuar group), Grand Canyon,

Arizona," *Journal of Paleontology* 59: 741–65; Porter S.M., R. Meisterfeld, and A.H. Knoll. 2003. "Vase-shaped microfossils from the Neoproterozoic Chuar group, Grand Canyon: a classification guided by modern testate amoebae," *Journal of Paleontology* 77: 409–29.

17. Bengtson S., B. Rasmussen, and B. Krapež. 2007. "The paleoproterozoic megascopic Stirling biota," *Paleobiology* 33: 351–81.

18. Rai V., and R. Gautam. 1999. "Evaluating evidence of ancient animals," *Science* 284: 1235a; Seilacher A., P.K. Bose, and F. Pflüger. 1999. "Evaluating evidence of ancient animals," *Science* 284: 1235a.

19. We distinguish different groups of fauna, usually named for the first locality in which they were discovered: the Ediacara fauna (565 to 543 million years ago), the SSF (small shelled fossils) fauna (543 to 530 million years ago), the Tommotian and Atdabanian faunas (530 to 521 million years ago), the fauna of Chengjiang (530 millions years ago), and the Burgess fauna (515 million years ago). But it is in 590–570-million-year-old rocks in Doushanto, China, that the first group of multicellular animals fossils were discovered. These are sponges or enigmatic balls of cells resembling embryonic stages.

20. The primary era is also called the Paleozoic. Cambrian comes from *Cambria,* the name the ancient Romans gave to present-day Wales. In this region in 1835, English geologist Adam Sedgwick (1785–1873) discovered sediments that were then believed to be the oldest yet discovered.

21. In one of his first books, Stephen J. Gould distinguished two possibilities (Gould, *Ever Since Darwin* [see chapter 14]).

22. The blackish imprints left by the carbon of these soft-bodied organisms, trapped between two layers of mud, are rarely found but are ancient vestiges of life, such as those of the oldest fossils of plants visible to the naked eye (*Grypania spiralis,* dating from 2.1 billion years ago) or of the extraordinary fauna of Burgess (515 million years ago).

23. Many scientific publications describe research on the date at which oxygen appeared on Earth and its increasing concentration between 3.8 and 0.57 billion years ago. Because oxygen is a gas that acts on many minerals, geologists have sought to date the first traces of oxidation. Banded iron formations (BIFs) date from 2.5 to 2.3 billion years ago. These iron oxides are deposited in marine habitats by precipitation in oxygenated water. Therefore, at the time the first banded iron formations were formed, oxygen was already present in the marine environment. The discovery of red layers of ferric salts in other rocks is explained by precipitation in continental waters; therefore oxygen was present in the atmosphere. These date from 2.3 billion years ago or later. Much other reasoning based on paleontology and geology allows us to establish that, 2.5 billion years ago, the concentration of oxygen dissolved in water was about 0.5 percent of the current concentration. This concentration increased slowly and reached between 7 percent and 10 percent of present-day concentrations 570 million years ago. There was truly too little oxygen to allow a precocious explosion of macroscopic life.

24. In his first collection of short articles, Stephen J. Gould devoted a chapter to the relationship between size and shape (Gould, *Ever Since Darwin* [see chapter 21]).

25. The cytopharynx of unicellular ciliates, a sort of funnel to which prey are directed, and where they are then taken in, is probably the fruit of a long evolutionary sequence, and this is why it appeared late.

26. Stephen J. Gould has developed the idea that the Cambrian explosion is only a false impression of suddenness in the elaboration of life visible to the naked eye. As for many apparently explosive phenomena in biology, there would have been a phase of slow evolution leaving few traces (Gould, *Ever Since Darwin* [see chapter 15]).

27. The following publication describes the divergence of the first multicellular animal lineage, dating to more than a billion years ago: Wray G.A., J.S. Levinton, and L.H. Shapiro. 1996. "Molecular evidence for deep Precambrian divergences among metazoan phyla," *Science* 274: 568–73.

28. The following publication challenges (27), estimating the first divergence at 670 million years ago: Ayala F.J., A. Rzhetsky, and F.J. Ayala. 1998. "Origin of the metazoan phyla: molecular clocks confirm paleontological estimates," *Proceedings of the National Academy of Sciences U S A* 95: 606–11. This affair led to a debate about the value of genetic interpretations and ways to limit errors in them. Publications on bias in the molecular clock technique include the following: Rodríguez-Trelles F., R. Tarrío, and F.J. Ayala. 2002. "A methodological bias toward overestimation of molecular evolutionary time scales," *Proceedings of the National Academy of Sciences U S A* 99: 8112–15; Sanderson M.J. 2002. "Estimating absolute rates of molecular evolution and divergence times: a penalized likelihood approach," *Molecular and Biological Evolution* 19: 101–2.

29. The first organisms to conquer land appeared about 470 million years ago in the Paleozoic era, between the Ordovician and the beginning of the Silurian. These were amphibious species, living in water as much as outside of it. The first traces of unidentified terrestrial plants date from 470 million years ago. They resembled spores of mosses (Gray J. 1993. "Major Paleozoic land plant evolutionary bio-events," *Palaeogeography, Palaeoclimatology, Palaeoecology* 104: 153–69). The first spores of "mosses," of hepatic bryophytes, date to 430 million years ago (Wellman C.H., P.L. Osterloff, and U. Mohiuddin. 2003. "Fragments of the earliest land plants," *Nature* 425: 282–85.) A molecular analysis shows that the hepatics and another group of "mosses," the anthocerotophytes, are ancestors of all terrestrial plants. See Nickrent D.L., C.L. Parkinson, J.D. Palmer, and R.J. Duff. 2000. "Multigene phylogeny of land plants with special reference to bryophytes and the earliest land plants," *Molecular Biology and Evolution* 17: 1885–95. Another molecular analysis indicates that terrestrial plants appeared 700 million years ago: Heckman D.S., D.M. Geiser, B.R. Eidell, R.L. Stauffer, N.L. Kardos, and S.B. Hedges. 2001. "Molecular evidence for the early colonization of land by fungi and plants," *Science* 293: 1129–33. The first terrestrial animals were probably arthropods (early insects derived from crustaceans). They left traces in an ancient aeolian dune dating to the beginning of the Ordovician (MacNaughton R.B., J.M. Cole, R.W. Dalrymple, S.J. Braddy, D.E.G. Briggs, and T.D. Lukie. 2002. "First steps on land: Arthropod trackways in Cambrian-Ordovician eolian sandstone, southeastern Ontario, Canada," *Geology* 30: 391–94). The first fossils of these arthropods (dating to the beginning of the Silurian) are of spiders (arachnids) or millipedes (myriapods) (Jeram A.J., P.A. Selden, and D. Edwards. 1990. "Land animals in the Silurian: arachnids and myriapods from Shropshire, England," *Science* 250: 658–61).

30. Tropical corals, red corals, soft corals, gorgonians, sea anemones, and jellyfish are the best-known representatives of the large group Coelenterata.

31. The birth of continents is poorly known before the supercontinent formation of Rodinia. One pertinent piece of evidence concerns a study of a 180-mile-wide meteorite impact crater in South Africa dating from 3.08 billion years ago, which suggests that the crustal layer formed first in an early continent (Moser E., R.M. Flowers, and R.J. Hart. 2001. "Birth of the Kaapvaal Tectosphere 3.08 billion years ago," *Science* 91: 465–68).

32. The main contribution to the theory of continental drift is attributed to Alfred Wegener (1880–1930). Understanding of this drift is attributed to a member of my family, the Dutch geophysicist Felix Andries Vening Meinesz (1887–1966), who contributed substantially to the study of convection of magma as part of the movement of tectonic plates.

33. The American geologist John L. Kirschvink first formulated a hypothesis of the Earth transformed into a snowball after a period of severe glaciation. Later, four episodes of glaciation of this type were identified and dated from between 750 and 600 million years ago and another to 2.4 billion years ago (Kirschvink J.L. 1992. "Late Proterozoic low-latitude global glaciation: the snowball Earth," in J.W. Schopf and C. Klein, eds., *The Proterozoic Biosphere* [New York: Cambridge University Press], 51–52; Kirschvink J.L., E.J. Gaidos, L.E. Bertani, N.J. Beukes, J. Gutzmer, L.N. Maepa, and R.E. Steinberger. 2000. "Paleoproterozoic snowball Earth: extreme climatic and geochemical global change and its biological consequences," *Proceedings of the National Academy of Sciences U S A* 97:1400–5; Hoffman P.F., A.J. Kaufman, G.P. Halverson, and D.P. Schrag. 1998. "A Neoproterozoic snowball Earth," *Science* 281: 1342–46).

34. Among the present-day unicellular algae, one group is dominant: diatoms or bacillariophytes (part of the Stramenopiles or Straminopiles). The current number of diatom species is about 20,000, but specialists estimate that their true diversity is more than 100,000 species, a number I have retained in my estimate of the diversity of chlorophyll-bearing protists (Round F.E., and R.M. Crawford. "Phylum Bacillariophyta," in L. Margulis, J.O. Corliss, M. Melkonian, and D.J. Chapman, eds., *Handbook of Protista* [Boston: Jones and Bartlett], 574–96). The other planktonic plants (about 10,000 species) are essentially parts of five evolutionary lineages: haptophytes or coccoliths, cryptophytes, euglenophytes, dinophytes or dinoflagellates, and chlorophytes.

35. The biodiversity of multicellular algae is better known. They belong primarily to three lineages. The green algae are part of the chlorophytes, red algae are rhodophytes, and brown algae are phaeophytes (part of the Straminopiles). Two recent synthetic works describe the diversity of unicellular and multicellular algae in all lineages: Hoek C. van den, D.G. Mann, and H.M. Jahns. 1995. *Algae: An Introduction to Phycology* (Cambridge: Cambridge University Press); Reviers B. de. 2003. *Biologie et phylogénie des algues* (Paris: Belin).

36. The macroscopic floating (pelagic) algae are less numerous. The sargassums of the north Atlantic, off the United States coast, are the best known. For flotation, these brown algae (phaeophytes) have aerocysts, balloon-like structures filled with gas.

37. Pliny the Elder originally said these words in the context of the smallest creatures visible to the naked eye. In 79 BC, he died across the Bay of Naples from Pompeii while attempting to rescue people from the eruption of Mount Vesuvius (Gould, *The Panda's Thumb* [see the prologue]).

38. Darwin, *On the Origin of Species* (see chapter 10).

39. The paleontologists J. William Schopf and Andrew H. Knoll have contributed greatly to the discovery of the vestiges of the flora and fauna living well before the Cambrian explosion of life (Schopf, J.W. 2000. "Solution to Darwin's dilemma: discovery of the missing Precambrian record of life," *Proceedings of the National Academy of Sciences U S A* 97: 6947–57; Knoll, *Life on a Young Planet*).

40. This metaphor by the founder of modern geology, Sir Charles Lyell (1797–1875), was taken up again by Darwin (Darwin, *On the Origin of Species* [see the last paragraph of chapter 10]).

Chapter 7

1. The largest alga in the world is *Macrocystis pyrifera* (a brown alga belonging to the Phaeophyceae, Straminopiles).

2. Representatives of microalgae called diatoms (also known as Bacillariophyceae, they are part of the stramenopiles), cryptophytes, or haptophytes (which include the coccolithophores) are often found fixed or buried in mud in a unicellular state. Similarly, many unicellular protozoans crawl over or are fixed to the seafloor (such as certain foraminiferans, rhizopods, and ciliates).

3. There are also garlands of cells stuck more or less firmly to one another in certain diatoms of the genus *Chaetoceros* or green algae of the genera *Scenedesmus* or *Zygnema*.

4. A representative of the haptophytes—*Corymbellus aureus*—can be either unicellular or composed of a colony of cells all while remaining subfloating. When living colonially, it can take various forms: a small hollow sphere formed of a layer of flagellated cells, a filled sphere made up of unflagellated cells, or a macroscopic, shapeless colony composed of tens of thousands of cells enmeshed in mucus (Van den Hoek, Mann, and Jahns, *Algae: An Introduction to Phycology* [229]). The green alga (Chlorophyta) *Palmophyllum crassum* consists of a colony of cells fixed to the seafloor. It is a crust of beautiful green that carpets the rocky slopes of the Mediterranean. Each crust is composed of thousands of cells stuck together with no predetermined order. Similarly, the red alga (Rhodophyceae) *Porphyridium purpureum* has the form of a crust composed of a multitude of cells joined within a mass of mucus.

5. It was Antoni van Leeuwenhoek who discovered this. He was the first to observe green balls composed of 500 to 60,000 cells joined together in a gel—the green alga *Volvox*. He was fascinated by this mobile, changing structure. Another remarkable organism illustrates an intermediate state: *Dictyostelium discoideum*. Classed among the slime molds or animal-like fungi (Mycetozoa), it is seen either in a unicellular amoeboid state or as an aggregate of cells. Colonies composed of a huge number of cells (nearly 100,000) take the form of crawling snail. The American professor John T. Bonner was excited by the origin of multicellularity (he wrote 15 books that develop this theme). He valued the peculiar form of *Dictyostelium discoideum,* precursor of multicellular life. See Bonner J.T. 1998. "A way of following individual cells in the migrating slugs of *Dictyostelium discoideum*," *Proceedings of the National Academy of Sciences U S A* 95: 9355–59; idem. 2000. *First Signals: The Evolution of Multicellular Development* (Princeton: Princeton University Press); Cox E.C., and J.T. Bonner. 2001. "The advantages of togetherness," *Science* 292: 448–49.

6. If we analyze the strategies of cellular attachment, we observe that they are similar to behaviors we see today in many plant and animal groups. Thus, the term *neutralism* is suitable if no exchange occurs between two neighboring cells, and if neither lives at the expense of the other. The term *commensalism* applies if one cell plays the role of an indifferent partner and the other cell clearly benefits by the association, which allows it to feed itself better (this term is especially used among animals, and if the neutral partner is a plant we call the partner that benefits an *epiphyte*). The term *adelphoparasite* is appropriate if the second cell grows because of substances transmitted by the first cell, but without conferring reciprocal benefits (this term was coined to designate a parasite feeding on the same species, or a closely related one; it also designates the mode of development of the carposporophyte on the female gametophyte of red algae). The term *multicellular organism* describes the case in which a symbiotic relationship forms between daughter cells. The two cells establish exchanges that are beneficial and indispensable to both. This is in fact a *symbiosis* between two generations of cells (parasitologists prefer the term *mutualism*). These are social pacts that are inscribed in the genes. Any wholly independent life is impossible from then on in the life cycle of these species. This is the initiation of cellular societies.

7. Many species of the green alga phylum (Chlorobionta), representing different orders, consist of just a simple, unbranched file of cells. Different strategies lead to this archaic

multicellular state. The formation of a wall perpendicular to the main axis of the mother cell is the most frequent strategy (*Ulothrix, Uronema, Klebsormidium*). In the genus *Cylindrocapsa,* the filament is formed by a succession of cell divisions. But after each division, the two daughter cells remain attached end to end inside the original cell wall (see fig. 20d). A filament is thereby produced that is constructed according to the nested matryoshka doll model.

8. This state, unicellular but multinucleated, is called syncytial, cœnocytic, or siphonous, depending on the taxon.

9. A thousand species of filamentous fungi, such as some stramenopiles (some phycomycetes including mildew), and some eumycetes (zygomycetes) are also cœnocytic. The amoeboid fungi (the myxomycetes or Mycetozoa) are represented by a thousand species; each mature amoeba can contain several nuclei.

10. These are certain ciliates (*infusoria*) and stages of development of microsporidia, unicellular parasites. Some of these multinucleated unicells can reach lengths of 3 millimeters.

11. The order Cladophorales (in the green algae, Chlorophyta) and many red algae (Rhodophyta) have multi-nucleated segments. This trait is also seen in certain animal cells, such as those of our muscles.

12. The genus *Grypania* has left traces in rocks dating from 2.1 billion years ago. Many fossils of this enigmatic genus have been found in various sites, incrusted in rocks dating to diverse epochs over the next billion years (Han and Runnegar, "Megascopic eukaryotic algae from the 2.1-billion-year-old Negaunee iron-formation, Michigan"). The order Dasycladales (siphonated green algae) arose in the Precambrian (many traces of various representatives of this order in the Ulvophyceae, more or less accepted by specialists, are dated from 1.2 to 0.6 billion years ago). Another alga of the genus *Vaucheria* is one of the first algae visible to the naked eye embedded in rocks (dated to 750 million years ago). The genus *Vaucheria* (Straminopiles, Xanthophyceae) is still found today. Its vegetative apparatus consists of filaments as fine as hairs, each made up of a single cell with many nuclei (Butterfield N. 2002. "A *Vaucheria*-like fossil from the Neoproterozoic of Spitsbergen," *Proceedings of the Denver annual meeting of the Geological Society of America* 75: 3).

13. The syncytial theory of metazoan evolution was first proposed by Jovan Hadzi in 1953. Stephen J. Gould also relates the same hypothesis, defended by Earl D. Hanson in 1958. According to Hanson, the simplest modern flatworms (platyhelminths), worms without an internal cavity (acoelomate), were formed by segmentation of multi-nucleated ciliates (See Gould, *The Panda's Thumb* [see chapter 24]; Hadzi J. 1953. "An attempt to reconstruct the system of animal classification," *Systematic Zoology* 2: 145–54; idem. 1963. *The Evolution of the Metazoa* [New York: Macmillan]; Hanson E.D. 1958. "On the origin of the Eumetazoa," *Systematic Zoology* 7: 16–47; idem. 1977. *The Origin and Early Evolution of Animals* [Middletown, CT: Wesleyan University Press]). The scenario of delayed cell division is actually common in some algae like in *Dictyosphaeria* (fig. 18a) and other Siphonocladales (green algae, Chlorophyta). In *Dictyosphaeria,* a giant cell, or vesicle, containing many nuclei is at the origin of a multicellular ball visible to the naked eye owing to a segmentation (segregative cell divisions) of the interior space (fig. 20a). See Enomoto S., T. Hori and K. Okuda. 1982. "Culture studies of *Dictyosphaeria* (Chlorophyceae, Siphonocladales). II. Morphological analysis of segregative cell division in *Dictyosphaeria cavernosa,*" *Japanese Journal of Phycology* 30: 103–12.

14. Geneticists have demonstrated that the flatworms are not platyhelminths but the simplest representatives of bilateral metazoans. Carranza S., J. Baguñà, and M. Riutort. 1997. "Are the Platyhelminthes a monophyletic primitive group? An assessment using 18s

rDNA sequences," *Molecular Biology and Evolution* 14: 485–97; Ruiz-Trillo I., M. Riutort, D.T.J. Littlewood, E.A Herniou, and J. Baguñà. 1999. "Acoel flatworms: earliest extant bilateral metazoans, not members of Platyhelminthes," *Science* 283:1 919–23.

15. It is among the green algae (Ulvophyceae, Chlorobionta) order Caulerpales that we find coenocytic representatives composed of a stolon creeping along the seafloor (which can reach three meters in length), fronds that closely resemble leaves of spermatophytes (seed plants), and bouquets of rhizoids that play a role similar to that of roots.

16. Butterfield et al., "A bangiophyte red alga from the Proterozoic of arctic Canada"; Butterfield N.J., "*Bangiomorpha pubescens* n. gen., n. sp.: implications for the evolution of sex, multicellularity and the Mesoproterozoic/Neoproterozoic radiation of Eucaryotes."

17. All the identical cells of the unbranched green alga *Ulothrix* (Ulvophyceae, Chlorobionta) can divide.

18. All cells of many species of green algae in the class Zygnematophyceae (Chlorobionta), which form an unbranched filament of cells, can transform into spores (genera *Mougeotia, Spyrogyra,* and *Zygnema*).

19. A unique apical growth is seen in filamentous red algae; sometimes they are branched, as in *Rhodochaete parvula* (rhodophytes). In brown algae, I can cite *Ectocarpus* (Phaeophyceae-Straminopiles), of which only the cells at the ends of certain filaments can become segmented, while other filaments have a specific "intercalary" zone in which cells are destined to divide (this is, in effect, a meristem).

20. It is of course to Edward O. Wilson's work that I am winking my eye here. In certain aspects of animal behavior (sociobiology), E.O. Wilson is by far the best known author because of his meticulous research and many publications, especially on ants. Among his important works are: Wilson E.O. 1975. *Sociobiology: The New Synthesis* (Cambridge, MA: Harvard University Press); Hölldobler B., and E.O. Wilson. 1994. *Journey to the Ants* (Cambridge, MA: Harvard University Press).

21. The Belgian biologist Christian de Duve has emphasized the benefits of collective action by cells (de Duve, *Vital Dust* [see chapter 18]).

22. Flattened forms are also found among the most archaic animals still in existence: acoelomate worms (lacking a body cavity, but shown by genetic analysis to be related to the Bilateralia) and minuscule crusts formed of two layers of cells separated by an interstice, such as the strange *Trichoplax adhaerens* (only member of the Placozoa).

23. Planktonic cellular colonies shaped like a hollow ball whose wall is composed of a layer of flagellated cells, seen notably in species of *Synura* (Chrysophyceae, Straminopiles), *Corymbellus aureus* (haptophytes), and *Chaetosphaeridium* (Klebsormidiophyceae, Chlorobionta).

24. These fossils have been found in only one region, in the southwest of China. See Xiao S., Y. Zhang, and A.H. Knoll. 1998. "Three-dimensional preservation of algae and animal embryos in a Neoproterozoic phosphorite," *Nature* 391: 553–58; Xiao S. 2002. "Mitotic topologies and mechanics of Neoproterozoic algae and animal embryos," *Paleobiology* 28: 244–50; Li C.-W., J.-Y. Chen, and T.-E. Hua. 1998. "Precambrian sponges with cellular structures," *Science* 279: 879–82; Chen, J.-Y., P. Oliveri, C.-W. Li, G.-Q. Zhou, F. Gao, J.W. Hagadorn, K.J. Peterson, and E.H. Davidson. 2000. "Precambrian animal diversity: putative phosphatized embryos from the Doushantuo formation of China," *Proceedings of the National Academy of Sciences U S A* 97:4457–62.

25. Stephen J. Gould presents two figures in which Haeckel's drawings are compared to images of the first fossils of multicellular animals (Gould S.J. 2000. *The Lying Stones of*

Marrakech: Penultimate Reflections in Natural History [New York: Harmony Books]; see chapter 21).

26. Phylogenesis (from the Greek *phulon,* meaning race and genesis) refers to the mode of formation of species, the evolutionary history of species, lineages, and groups of related organisms. Ontogenesis (from the Greek *onto,* meaning to be, being, and genesis) refers to the development of an individual, from a fertilized egg to adulthood (the term *ontogeny* is also used).

27. The morula stage is a form that arises from the first division in which all cells are still joined.

28. The blastula stage is a hollow ball formed of a single cell layer; the cavity is called a blastocoel.

29. The primitive multicellular filament of bryophytes is called a protonema.

30. Stephen J. Gould devoted his first book to the controversial theme of the similarity between ontogeny and phylogeny (Gould S.J. 1977. *Ontogeny and Phylogeny* [Cambridge, MA: Harvard University Press]).

31. The resemblance of choanoflagellates to nurse cells in sponges (choanocytes) was established over a century ago. Today there are about 150 of these protist species. It is also noteworthy that the forms of colonies of unicellular algae in various phyla greatly resemble choanocyte colonies. I can cite *Synura* (Chrysophyceae, Straminopiles), *Corymbellus* (haptophytes), and *Chaetosphaeridium* (Kelbsormidiophyceae, Chlorobionta), whose flagella are found at the base of a collar.

32. Studies on proteins and genes show a strong relationship between choanoflagellates and sponges. King N., and S.B. Carroll. 2001. "A receptor tyrosine kinase from choanoflagellates: molecular insights into early animal evolution," *Proceedings of the National Academy of Sciences U S A* 98: 15032–37; Snell E.A., R.F. Furlong, and P.W. Holland. 2001. "Hsp 70 sequences indicate that choanoflagellates are closely related to animals," *Current Biology* 11: 967–70.

33. The choice of shapes is definitely oriented by multiple physical constraints (such as gravity and surface tension) and chemical ones (as in determining protein shape). The English biologist D'Arcy Wentworth Thompson (1860–1948) founded the school of theoreticians who saw a crucial role for these constraints in the elaboration of life (to the point that they sometimes de-emphasized the role of natural selection and random mutations) (Thompson, *On Growth and Form*).

34. John Maynard Smith (1920–2004) described the development of multicellular species as modular. According to him, this is the mode that allows continual, gradual evolution. Evolution changes the last stages in the growth of modules, leaving their initial arrangement intact (Maynard Smith J. 1998. *Shaping Life: Genes, Embryos and Evolution* [London: Weidenfeld & Nicholson]).

35. The very first cells of an embryo are undifferentiated. They are said to be *totipotent,* that is, they retain all the potential of evolution, and they are currently the object of much applied research; the cloning of these cells could lead to grafts to reconstitute any cellular tissue.

36. King and Wilson, "Evolution at two levels in humans and chimpanzees"; The chimpanzee sequency and analysis consortium, "Initial sequence of the chimpanzee genome and comparison with the human genome"; Li and Saunders, "The chimpanzee and us."

37. A consortium of more than 200 authors published the first sequencing of the mouse genome (Mouse genome sequencing consortium and mouse genome analysis group.

2000. "Initial sequencing and comparative analysis of the mouse genome," *Nature* 420: 520–62).

38. Stephen J. Gould developed the principle that evolution moves inevitably towards increasing complexity as a logical result of the pathways open to evolution, rather than as an active tendency (Gould, *Life's Grandeur* [see chapter 14]).

39. At the beginning of chapter 2, I described the absence of accepted knowledge about the existence of living organisms simpler than bacteria (aside from viruses).

40. "Unity is strength"—in French, "L'union fait la force"—is the national motto of Belgium, where people who speak French and others who speak Flemish are linked in this same nation. Union has always conferred strength in civilizations throughout history. The term occurs in names of nations or national emblems, such as *United States*, *United Kingdom*, *European Union*, or the flag *Union Jack*.

41. The six lineages of plants with chlorophyll that have multicellular species are the green plants (including the green algae Chlorophyceae or Ulvophyceae, which are part of the lineages of chlorophytes or Chlorobionta, which include all green plants), red algae (rhodophytes or Rhodobionta), brown algae (essentially the Phaeophyceae, which are part of the Straminopiles), and three lineages in which multicellular species are in the minority. These three lineages are the haptophytes (primarily unicellular, though certain species, such as those in the genus *Pleurochrysis,* have multicellular stages), cryptophytes (unicellular algae except for the multicellular *Bjornbergiella*), and dinophytes (algae that are unicellular except for a few species that have multicellular stages, such as those in the genus *Dinoclonium*).

42. The main classes of fungi are multicellular and belong to one lineage: the eumycetes (including the ascomycetes and the basidiomycetes, which together form the septate fungi, the septomycetes). All classes of multicellular animals make up the lineage of Metazoa.

43. The elaboration of forms (morphogenesis) is a subject frequently discussed in biology. The German paleontologist Adolf Seilacher defined three groups of factors that can orient the construction of organisms: genetics (the historical capital of the species), structural constraints (the ensemble of physical and chemical forces that orient forms towards logical architectures), and adaptation to the environment. Debates arise about the preeminence of one axis over the two others. Seilacher views them as equally important: this is his triangle theory (Seilacher A. 1970. "Arbeitskonzept zur konstruktionsmorphologie," *Lethaia* 3: 393–96).

44. Gould, *The Panda's Thumb* (see chapter 3).

45. The importance of constraints that structure living organisms was initially propounded by D'Arcy W. Thompson in 1917 (Thompson, *On Growth and Form*). The theme of convergences as a tangible result of the importance of structural constraints limiting the pathways chance can lead to is argued by Simon Conway Morris in two works (Conway Morris S. 1998. *The Crucible of Creation: The Burgess Shale and the Rise of Animals* [Oxford: Oxford University press]; idem. 2003. *Life's Solution; Inevitable Humans in a Lonely Universe* [Cambridge: Cambridge University Press]).

46. de Duve, *Singularities*.

47. Spinoza wrote *Ethics* beginning in 1662, but this work was not published until after his death in 1677. These citations are taken from the appendix of the first chapter, "Concerning God."

48. Stephen J. Gould cites the works of the paleontologist Louis Dollo. According to him, an elementary calculation of probabilities demonstrates the impossibility for convergence ever to approach perfect resemblance. Gould, *The Panda's Thumb*.

49. The cyanobacteria, long classed among the algae (the Cyanophyceae), are photosynthetic bacteria possessing chlorophyll. The multicellular cyanobacteria constitute the present-day stromatolites and are likely the basis for the origin of the fossil stromatolites (Schopf J.W. 2000. "The paleobiologic record of cyanobacterial evolution," in Y.V. Brun and L.J. Shimkets, eds., *Prokaryotic Development* [Washington, DC: ASM Press], 105–29).

50. The different multicellular cyanobacteria are classified according to their degree of complexity (see fig. 5 in chapter 3). We thus distinguish the Oscillatoriales (filaments of undifferentiated cells with false branchings), Nostocales (unbranched filaments with cellular differentiation: the akinetes and the heterocysts), and Stigonematales (filaments that are sometimes branched and even arranged in several rows with cellular differentiation: the akinetes and the heterocysts). "The earliest known akinetes are preserved in ≈ 2,100-Ma [million years ago] chert from West Africa. Geochemical evidence suggests that oxygen first reached levels that would compromise nitrogen fixation and hence select for heterocyst differentiation 2,450–2,320 Ma. Integrating phylogenetic analyses and geological data, this suggests that the clade of cyanobacteria marked by cell differentiation diverged once between 2,450 and 2,100 Ma" (Tomitani A., A.H. Knoll, C.M. Cavanaugh, and T. Ohno. 2006. "The evolutionary diversification of Cyanobacteria. Molecular–phylogenetic and paleontological perspectives," *Proceedings of the National Academy of Sciences U S A* 103: 5442–47).

51. One of the main theoreticians of bacterial evolution, W. Ford Doolittle, has presented a diagram of a great merging among the original bacterial lineages. Before the first endosymbioses, there were many gene exchanges between the Archaea and the Eubacteria (Doolittle W.F. 1999. "Phylogenetic classification and the universal tree," *Science* 284: 2124–28).

52. John Maynard Smith depicted the programmed, modular development of species. To explain this viewpoint, he invoked the series of Hox genes that are activated one after the other in the course of development of the mouse but also of the fly. This kind of structuring programming exists in all animals (Maynard Smith, *Shaping Life*). Moreover, it has been observed that the absence of a protein in the colonial amoebas of *Dictyostelium* (see note 5 above) prevents the formation of a highly structured stalk in colonies of this species. The gene corresponding to this protein has been identified (Ennis H.L., D.N. Dao, S.U. Pukatzki, and R.H. Kessin. 2000. "*Dictyostelium* amoebae lacking an F-box protein form spores rather than stalk in chimeras with wild type," *Proceedings of the National Academy of Sciences U S A* 97: 3292–97).

53. To describe the two materialist approaches to the evolution of life, John Maynard Smith distinguished the reductionist viewpoint to describe information-processing, the regulation and control of genetically programmed development, from the holistic (systemic) viewpoint that sees a dynamic, mechanical self-organization of life, essentially oriented by unbreakable laws of chemistry and physics (Maynard Smith, *Shaping Life*).

54. A spiral composed of plant and animal forms representing biodiversity is the emblem of French national parks.

Chapter 8

1. *Candide, ou l'optimisme* is a philosophical tale published in 1759. Its author, François-Marie Arouet, *alias* Voltaire, was then 64 years old (Voltaire. 1759. *Candide, ou l'optimisme, traduit de l'allemand de Mr le Docteur Ralph* [Geneva: Cramer]; American edition: 2000. *Candide and Related Texts* [Indianapolis: Hackett]).

2. Voltaire reignited the old philosophical debate between disciples of Aristotle, who believed in predestined causes, and those of Lucretius, who favored progressive adaptation that cannot be predicted initially. Voltaire wrote *Candide* to criticize the German philosopher Gottfried Wilhelm Leibniz (1646–1716), who had just developed arguments in favor of predestined causes.

3. Voltaire, *Candide* (quote from p. 2). Since his appearance, biologists have used the caricature of Pangloss to denounce finalism, the theory that tends to justify a finality in nature and that is attributed to Aristotle, Liebniz, and Arthur Schopenhauer (1788–1860). It stipulates that God created all beings according to his own goals, everything is predetermined. For example, the nose was created to hold up glasses. By extension, all descriptions that see evolution tied to a particular adaptation are called panglossian; this view takes no account of the multiple redirecting of natural selection, which continually creates opportunistic secondary adaptations to functions with multiple constraints.

4. "Tout va très bien Madame la Marquise" ("Everything is OK, Madam") is a famous French song written by Paul Misraki, a member of the Ray Ventura Orchestra (© 1936 Ventura/Tutti/Warner Chappell).

5. Stoicism is a philosophy created in Athens by Zeno of Citium (336–264 BC). Stoics do not struggle against fate; they always welcome it. The Roman Seneca (43–56 AD) was one of the best-known disciples of stoicism. In accord with his philosophical stance, he committed suicide when ordered to by the Emperor. Like Pangloss, Seneca was a tutor (his pupil was the emperor Nero).

6. The term *dinosaur* was first proposed in 1842 by the English paleontologist Richard Owen (1804–1892). It means "terrifying lizard" and designates the first group of fossil reptiles discovered (*Megalosaurus* and *Iguanodon*).

7. It was hypothesized in the 1950s that dinosaur sperm were denatured in internal testicles, overheated by the metabolism of these monsters. In this view, the mammals survived because their testicles were outside their bodies. Of course, this is a crackpot idea.

8. At first, the meteorite hypothesis was primarily advocated by two related scientists: the father, physicist, and Nobel laureate Luiz W. Alvarez (1911–1988), and the son, geologist Walter Alvarez. See Alvarez L.W., W. Alvarez, F. Asaro, and H.V. Michel. 1980. "Extraterrestrial cause for the Cretaceous-Tertiary extinction: experimental results and theoretical interpretation," *Science* 208: 1095–108; Alvarez W. 1997. T. Rex *and the Crater of Doom* (Princeton, NJ: Princeton University Press). Their hypothesis was initially highly controversial. The polemics and scientific debates surrounding it are summarized and analyzed in a collective work. See Glen W., ed. 1994. *The Mass-Extinction Debates: How Science Works in a Crisis* (Stanford, CA: Stanford University Press).

9. In 1946, Georges Remi (1907–1983) *alias* Hergé (phonetic pronunciation of the initials of his name) published the comic strip "L'étoile mystérieuse," in which a meteorite crashes into the sea, causing no damage but liberating giant organisms (Hergé. 1946. *Les aventures de Tintin et Milou. L'étoile mystérieuse* [Tournai: Casterman]; American edition: *The Adventures of Tintin: The Shooting Star* [Boston: Little, Brown, 1978]).

10. A geologist at Virginia Tech University, Dewey McLean, first established a link between the basalts of Deccan and the Cretaceous-Tertiary mass extinctions. But it was above all the French geologist Vincent E. Courtillot who proposed this alternative explanation. He wrote a book on the battle he fought to get the scientific community to consider the role of volcanic eruptions in mass extinctions. See McLean D. M. 1982. "Deccan volcanism and the Cretaceous-Tertiary transition scenario : a unifying causal mechanism," *Syllogeus* 39: 143–44; Courtillot V.E. 1986. "What caused the mass extinction? A volcanic eruption,"

Scientific American 263: 85–92; idem. 1999. *Evolutionary Catastrophes: The Science of Mass Extinction* (Cambridge: Cambridge University Press).

11. Deccan is a province in western India, and "traps" means step of a staircase in several northern European languages (including Dutch and Swedish). The Deccan Traps are stacked basaltic lava flows that give the mountains a finely striped appearance. The lava layer reaches 2,000 meters in height and extends over more than 500,000 square kilometers. One of my colleagues at the University of Nice–Sophia Antipolis, Gilbert Féraud, proposed a precise dating of these formations by using argon isotopes; according to him, the titanic successive regurgitations of Deccan lava date from 63 to 67 million years ago.

12. Baron Jean L.N.F. Cuvier (Georges Cuvier) was the first to provide a theory according to which catastrophes (global revolutions) periodically destroy the living world: "The species of ancient times were as unvarying as ours, or was it because the catastrophe that destroyed them did not give them time to produce variation? Numberless living beings have been victims of these catastrophes; those inhabiting dry land were engulfed by floods; others, who peopled the heart of the oceans, were desiccated when the seafloor suddenly rose; their races were terminated forever, and left in the world only some debris that is barely recognizable to the naturalist" (Cuvier G. 1815. *Discours sur les révolutions de la surface du globe, et sur les changemens qu'elles ont produits dans le règne animal.* 3rd ed. [Paris: Dufour et D'Ocagne]; the first edition was published in 1812 as a "discourse préliminaire" of the work, *Recherches sur les ossemens fossilles de quadrupèdes: où l'on rétablit les caractères de plusieurs espèces d'animaux que les révolutions du globe paroissent avoir détruites* [Paris: Déterville]. Translation by Ian Johnston available online at http://www.mala.bc.ca/~Ejohnstoi/cuvier.htm).

13. The three following publications provide data suggesting several very old, severe periods of glaciation: Kirschvink, "Late Proterozoic low-latitude global glaciation: the snowball Earth," in Schopf and Klein, eds., *The Proterozoic Biosphere*, 51–52; Kirschvink et al., "Paleoproterozoic snowball Earth: extreme climatic and geochemical global change and its biological consequences"; Hoffman et al., "A Neoproterozoic snowball Earth."

14. These fossils were first found in 1947 in the Ediacara Hills, located 650 kilometers north of Adelaide in South Australia. These vestiges are evidence of a marine fauna that lived between 565 and 543 million years ago. Since then, many similar forms of soft, fossilized animals dating from the same epoch have been exhumed on all continents.

15. The Pennatulacea are soft corals. I saw modern representatives of these filter-feeders near Nice at a depth of about 50 meters. They resemble feathers anchored in the marine substrate by an appendix shaped like a large radish.

16. The Tommotian fauna mark an important stage in animal evolution. Unlike the remains of animals of the Ediacaran period, they have calcified parts (these are the parts that have been well preserved and discovered). The fossils of these animals date from 530 million years ago and have been found at many sites worldwide. But their traces are absent from rocks dated to several million years later. No one knows what the minuscule calcareous strips, needles, or cups of this fauna contained or what supported them.

17. Hou X.-G., R.J. Aldridge, I. Bergström, D. Siveter, D. Siveter, and F. Xiang-Hong. 2004. *The Cambrian Fossils of Chengjiang, China: The Flowering of Early Animal Life* (Oxford: Blackwell).

18. The Burgess fossils are often represented by dark imprints (remains of carboniferous organic matter). The soft parts of these animals have therefore left traces that allow us to reconstruct their general form.

19. *Pikaia gracilens* was first described (in 1910) by Walcott as being a simple annelid worm in the form of a thick ribbon 5 centimeters long. In the late 1980s, Simon Conway

Morris reexamined all fossil specimens of *Pikaia* and concluded they were doubtless members of the chordates, that is, animals with a backbone and spinal cord. They are therefore the oldest representatives of the vertebrate lineage (to which we belong). Since then, other archaic fossil chordates similar to *Pikaia* have been found in China in rocks dated to more than 500 million years ago. See Chen J.-Y., J. Dzik, G.D. Edgecombe, L. Ramsköld, and G.-Q. Zhou. 1995. "A possible early Cambrian chordate," *Nature* 377: 720–22; Shu D.-G., S. Conway Morris, and X.-L. Zhang. 1996. "A *Pikaia*-like chordate from the lower Cambrian of China," *Nature* 384: 157–58.

20. In the 1970s and 1980s Harry Whittington and Simon Conway Morris published several dozen articles describing a different classification of the fossilized Burgess species than that established by Walcott.

21. Even Stephen J. Gould, to his regret, was caught in the trap of uncertainty about the ancestral forms of life. In his book, *Wonderful Life,* published in 1989, he discussed at great length the extraordinary diversity of the fossil fauna in the Burgess schists. Among the unusual animals cited, *Hallucigenia* was prominent. It was so dissimilar to any other known body plan that he hypothesized that it might represent a piece of another animal. Three years later, in another book published in 1993, *Eight Little Piggies,* Gould told how he learned to turn over *Hallucigenia* (all previous interpretations had pictured the animal upside down), at which point its form resembled that of a series of fossil and living species of onychophorans, a group located phylogenetically between the worms and the arthropods (Gould, *Wonderful life*; Gould, *Eight Little Piggies* [see chapter 24]). This mistake was also extensively discussed by Simon Conway Morris, the main researcher who studied the Burgess fauna. In a book primarily devoted to his work on these animals, he pointed to another error of interpretation, concerning *Anomalocaris*, which was originally identified as two separate fossils. Independently of these errors, Conway Morris propounded a view very different from that of Gould on the novelty of certain animal forms fossilized at Burgess (Gould saw them as extinct, Conway Morris as transitory) (Conway Morris, *The Crucible of Creation*).

22. The genus *Opabinia* is the strangest of the fossilized Burgess species. Long considered an archaic arthropod, it is now seen as belonging to no known animal lineage.

23. Gould, *Wonderful life*. See also his historical research on Walter D. Walcott: Gould, *Eight Little Piggies* (chapter 15).

24. Olsen P.E., D.V. Kent, H.-D. Sues, C. Koeberl, H. Huber, A. Montanari, E.C. Rainforth, S.J. Fowell, M.J. Szajna, and B.W. Hartline. 2002. "Ascent of dinosaurs linked to an iridium anomaly at the Triassic-Jurassic boundary," *Science* 296: 1305–7.

25. David M. Raup, an American paleontologist, is the one who has most stressed the importance of extinction as a force opposing speciation. According to him, to ignore mass extinction is analogous to studying demography without accounting for mortality. See Raup D.M. 1995. "The role of extinction in evolution," in Fitch and Ayala, eds., *Tempo and Mode in Evolution*, 109–24; Raup, *Extinction: Bad Genes or Bad Luck?*

26. In the Bible, the story of Noah is mentioned in the book of Genesis, chapters 5 through 9. In the Torah, the same story is in Genesis 6:9–11:32. In the Koran, the story of Noah (Nuh) is mentioned in the 28 versets of Sourate number 71.

27. In 1997, a team of American geologists led by William B. F. Ryan and William C. Pitman published their interpretation of the flood, adducing scientific data supporting their hypothesis. In their view, the story of the flood refers to a truly gigantic inundation caused by the sudden pouring of Mediterranean waters into the Black Sea (which was, at the time, only a freshwater lake below the level of the Mediterranean): Ryan

W.B.F, W.C. Pitman III, L.O. Major, K. Shimkus, V. Moskalenko, G.A. Jones, P. Dimitrov, N. Görür, M. Sakinç, and H.Yüce. 1997. "An abrupt drowning of the Black Sea shelf," *Marine Geology* 138: 119–26; Ryan W.B.F, and W.C. Pitman. 1999. *Noah's Flood: The New Scientific Discoveries about the Event that Changed History* (New York: Simon and Schuster).

28. William Ryan and his colleagues estimate that it took two years for the Black Sea to fill (see note 27), but more accurate calculations suggest 10 years: Myers P., C. Wielki, S.B. Goldstein, and E.J. Rohling. 2003. "Hydraulic calculations of postglacial connections between the Mediterranean and the Black Sea," *Marine Geology* 201: 253–67.

29. Several studies propose a hypothesis opposite to that developed by William Ryan and his colleagues (see note 27): this is that the Black Sea poured into the Mediterranean via the Sea of Marmara. There remains great disagreement about the circumstances and date of the event. See Aksu A.E., R.N. Hiscott, and D. Yaşar. 1999. "Oscillating quaternary water levels of the Marmara Sea and vigorous outflow into the Aegean Sea from the Marmara Sea–Black Sea drainage corridor," *Marine Geology* 153: 275–302; Kerey E., E. Meriç, C. Tunoğlu, G. Kelling, R.L. Brenner, and A.U. Doğan. 2004. "Black Sea–Marmara Sea quaternary connections: new data from the Bosphorus, Istanbul, Turkey," *Paleogeography, Paleoclimatology, Paleoecology* 204: 277–95.

30. In 1812, the French paleontologist Georges Cuvier criticized the old idea of the Black Sea pouring into the Mediterranean via the Sea of Marmara in these terms: "One has also wanted to attribute the supposed shrinking of the Black Sea and of the Sea of Azov to the rupture of the Straits of Bosporus that was supposed to have taken place at the assumed time of the flood during the reign of Deucalion. . . . If the Black Sea had been so high as is supposed, it would have found several paths to flow through the passes and lower plains than the current edges of the Bosporus. . . . What would have fallen suddenly one day in a cascade through this new passage, the small quantity of water that would have been able to flow at one time through such a narrow opening. . . . would have spread over the immense extent of the Mediterranean without causing a tidal wave of several toises [a toise is approximately six and one-half feet]." (*Discours sur les révolutions de la surface du globe, et sur les changemens qu'elles ont produits dans le règne animal* [see note 12 in this chapter]).

31. Historians have established that the legend of Noah was transmitted by the Mesopotamians to the Jews. The biblical legend dating from 2,900 years ago was inspired by the *Epic of Gilgamesh,* itself a late version of the *Atrahesis Epic* (written in Akkadian), dated from about 3,700 years ago (according to a cuneiform tablet from Nippur).

32. Voltaire. 1769. *Dictionnaire philosophique.* 5th ed. (New edition, 1964. Paris: Garnier-Flammarion; American edition, *Philosophical Dictionary* [New York: Basic Books, 1962]; see the article on the flood, quoted material from pp. 327–28).

33. Gould, *Ever Since Darwin.* (see chapter 17; quote from p. 172).

34. Stephen J. Gould develops Whiston's astounding catastrophist theories in Gould S.J. 1991. *Bully for Brontosaurus: Reflections in Natural History* (New York: W. W. Norton; see chapter 25).

35. Judeo-Christian and Muslim doctrines consider explanations of the origin of life as reported by the first prophets to be the basis for knowledge. Although it is generally admitted by theologians that such knowledge should be developed, it must always be based on holy scripture. This is the origin of the persistent image of a morally transcendent God the Creator. For Spinoza, it was important to separate the two concepts: one should not confound a commandment with something that one is trying to understand, that is, obedience with knowledge.

36. In the biblical Genesis, God created man in His image with the capacity to dominate life on earth (Genesis 26). But when Adam and Eve ate the apple, the forbidden fruit of the tree of knowledge of good and evil, the Eternal reproached them for having thus acquired evil knowledge; this is why they became mortal.

37. In a plea to associate scientific knowledge of evolution with religious faith, the American geneticist Robert Pollack found similarity between religious revelations and great scientific illuminations. Both are irrational, unpredictable, and rare, and they emerge from the unknown (of which God is part) (Pollack R. 2000. *The Faith of Biology and the Biology of Faith* [New York: Columbia University Press]). Likewise, the American biologist Kenneth R. Miller presented his way of seeing the divine spirit in evolution (he claimed his faith in Darwin's God). (Miller, *Finding Darwin's God*).

38. Leakey L., and R. Lewin. 1995. *The Sixth Extinction: Pattern of Life and the Future of Humankind* (New York: Doubleday).

39. My reading the book by Darwinian biologist and believer Kenneth R. Miller and two texts by Pope John Paul II (the encyclical letter "Fides et ratio" [Faith and reason] sent in 1996 to bishops, and "The church and evolution" [message of October 22, 1996 from Pope John Paul II to the Pontifical Academy of Sciences assembled in plenary session]) inspired this paragraph. See Miller, *Finding Darwin's God*.

40. The declarations of Rev. George Coyne and Pope Benedict XVI were reported by the Associated Press in a dispatch titled "Vatican's chief astronomer says 'intelligent design' doesn't belong in science class," dated November 18, 2005.

41. The paleontologist Simon Conway Morris has provided support for the existence of convergence in evolution. In his view, such convergence tends to show that there is more determinism and less chance in evolution. According to him, the evolution of life was thus channeled inevitably towards intelligent beings (Conway Morris, *Life's Solution*).

Chapter 9

1. Such a clock face is pictured in a book by J. William Schopf, *Cradle of Life*. A similar comparison is also used, reducing evolutionary time to a yearly calendar. On this scale, the full calendar corresponds to 4.5 billion years and a day represents 12.3 million years. Bacteria therefore appear in late March, algae in July, multicellular organisms in November, vertebrates at the end of November, mammals in mid-December, primates December 27, and finally, around 5:00 PM on December 31, the ancestors of man—and man himself around 9:00 PM.

2. Lynn Margulis and her son Dorion Sagan have imagined an evolutionary string in the form of a line at the bottom of a page that continues for several pages with captioned cross strokes representing the best known evolutionary landmarks (Margulis L., and D. Sagan. 1995. *What Is Life?* [New York: Simon & Schuster; repr. 2000, Berkeley: University of California Press]; see chapter 3).

3. Perceptions of sagittal time and cyclical time in different ages and cultures have been brilliantly analyzed by Stephen J. Gould (Gould S.J. 1987. *Time's Arrow, Time's Cycle: Myth and Metaphor in the Discovery of Geological Time* [Cambridge, MA: Harvard University Press]).

4. Teilhard de Chardin described these evolutionary pathways as parallel *orthogeneses*.

5. Seilacher, "Arbeitskonzept zur konstruktionsmorphologie."

6. Stephen J. Gould discussed at length the relative importance of the three poles of Seilacher's triangle, which Gould called the *aptive* triangle (Gould, *The Structure of Evolutionary Theory* [see chapter 10]).

7. These are constraints elucidated by D'Arcy W. Thompson in 1917 (Thompson, *On Growth and Form*).

8. Gould S.J., and R.C. Lewontin. 1979. "The spandrels of San Marco and the Panglossian paradigm: a critique of the adaptationist programme," *Proceedings of the Royal Society of London B* 205: 581–98. Gould was very proud of this major contribution to the understanding of evolution. He discussed it at length in his last, voluminous work. See Gould, *The Structure of Evolutionary Theory* (see chapter 11).

9. A spandrel is an unavoidable architectural element: it is the space located above a particular kind of opening. A pendentive is a spandrel located between two contiguous arches.

10. This semantic detail moved Gould to reply to his critics, to comment on the original publication, and to propose the word *spandrel* for any structure with predictable form but that appears in secondary fashion as a consequence of another selected evolutionary pathway. See Gould, S.J. 1997. "The exaptive excellence of spandrels as a term and prototype," *Proceedings of the National Academy of Sciences U S A* 94: 10750–55.

11. Stephen J. Gould was astounded at the virulence of these arguments, proof for him that the example of spandrels was useful in generating progress in understanding (Gould, "The exaptive excellence of spandrels as a term and prototype").

12. Having visited the basilica, I do not understand why the learned Stephen J. Gould, a connoisseur of vocabulary, referred to the basilica as a cathedral (a basilica is a sanctuary; the one in Venice holds the remains of the evangelist Mark). Similarly, I observed that what he calls a *narthex* in one of his last books (or vestibule in Catholic churches) is described on site (at Venice) as the *atrium* (passage between the square and the church). See Gould, *I Have Landed* (see chapter 20).

13. Gould collaborated with his colleague, Elisabeth Vrba, to propose this neologism for changes in function in the course of evolution. Among *aptations* (from *aptus,* meaning usage), he distinguished *adaptations* (from *ad* meaning for and *aptus* meaning usage, which refer to evolution by selection to improve a pre-existing usage) and *exaptations* (which are "co-opted" evolutionary changes and are therefore not selected for the use they were put to; the use was subsequently acquired): Gould S.J., and E.S. Vrba. 1982. "'Exaptation,' a missing term in the science of form," *Paleobiology* 8: 4–15.

14. This is the natural *extinction* of species.

15. This is *anagenesis*.

16. This is *cladogenesis*. The paleontologist and zoologist George Gaylord Simpson conducted the first detailed analysis of the diverse temporal modes of speciation (Simpson, *Tempo and Mode in Evolution*).

17. Sepkoski J.J. 1997. "Biodiversity, past, present, and future," *Journal of Paleontology* 71: 533–39.

18. George Gaylord Simpson, in his book *Tempo and Mode in Evolution,* distinguished three tempos of evolution, determined by different degrees of persistence of fossil forms:

- *trachytelic* species are quickly changing species (that last only a million years on average).
- *horotelic* species are those that remain relatively unchanged for 10 million years on average.
- *bradytelic* species are those of great longevity (100 million years on average). In this group are the "living fossils," such as the famous coelacanth, the fish with proto-legs.

19. The exceptional longevity of forms of cyanobacteria was unknown when George Simpson distinguished these three categories of species according to their evolutionary rates (see preceding note). J. William Schopf therefore proposed a new term for organisms that seem virtually eternal (same forms after 3.5 billion years): *hypobradytelic*. See Schopf, *Cradle of Life*.

20. I have often been surprised by the power of certain metaphors or literary citations used (and repeated) by scientists to describe a characteristic of life. In this chapter I have cited the metaphor of the triangular spaces of the basilica of San Marco in Venice. In this chapter and the preceding one, I cited Candide's tutor, Pangloss. These are caricatures of the idea of predestined goals (teleology) used to ridicule goal-oriented theories in biology, such as the one stipulating that organs have evolved from scratch to reach a predestined end (the eye thus evolved so we could see, the nose to support eyeglasses, and so forth). Another metaphor, among the most widely employed, is that of Alice and the Red Queen. These are characters in *Through the Looking-Glass* by Lewis Carroll, (the nom de plume of Charles Lutwidge Dodgson; 1832–1898). The Red Queen, holding little Alice by the hand, must keep moving at great speed in order just to stay in the same place. Alice is thereby entrained in a frantic race and sees with astonishment that nothing moves around her. In the land Alice finds herself in, one must not stop running. The paleontologist Leigh Van Valen used this image to illustrate the results of his research on the survival times of various lineages of fossil animals: he discovered that species succeed one another at a constant rate. Each species therefore seems constrained to evolve in order to adapt to the competition and to changes in habitat. In order to avoid being marginalized and then eliminated, the species must engage, like Alice, in a mad race—that of evolution. See Van Valen L. 1973. "A new evolutionary law," *Evolutionary Theory* 1: 1–30; Carroll L. 2001. *Through the Looking-Glass* (New York: Bloomsbury).

21. This is the theory of the *hopeful monster* formulated by Richard B. Goldschmidt (1878–1958): Goldschmidt R.B. 1940. *The Material Basis of Evolution* (New Haven, CT: Yale University Press).

22. This is the *punctuated equilibrium* theory proposed by Niles Eldredge and Stephen J. Gould. Eldredge N., and S.J. Gould. 1972. "Punctuated equilibria: an alternative to phyletic gradualism," in T.J.M. Schopf, ed., *Models in Paleobiology* (San Francisco: Freeman, Cooper & Co), 82–115.

23. The punctuated equilibrium theory does not apply to all species. It has been hotly debated by evolutionists. Before his death, Stephen J. Gould defended his theory at length in his synthesis of evolution, *The Structure of Evolutionary Theory*.

24. Leakey and Lewin, *The Sixth Extinction*.

25. I owe this beautiful expression to Teilhard de Chardin. See Teilhard de Chardin, *Le phénomène humain*.

26. In 1989, Professor Robert T. Paine won the ECI Prize of the International Ecology Institute. This is perhaps the most prestigious among the few distinctions accorded ecologists. Alfred Nobel (born 1833, died in 1896 in San Remo, an Italian city 30 kilometers from Nice) endowed in his will a foundation to distribute prizes to representatives of various disciplines, but there is no Nobel Prize in ecology, a field that barely existed when Nobel died. The ECI Prize was accompanied by the publication of a book in a series entitled "Excellence in Ecology," summarizing the research of the recipient (Paine R.T. 1994. *Marine Rocky Shores and Community Ecology: An Experimentalist's Perspective*, Excellence in Ecology 4 [Oldendorf/Luhe, Germany: Ecology Institute]).

27. A book richly illustrated with photographs of the marine flora and fauna of Tatoosh Island was published by Anne Wertheim Rosenfeld and Robert T. Paine: Wertheim Rosenfeld A., and R.T. Paine. 2002. *The Intertidal Wilderness* (Berkeley: University of California Press).

28. The hydraulic model of the Seto Inland Sea is located 20 kilometers north of Hiroshima, in Kure at the Chugoku National Industrial Research Institute. The model was constructed in 1973 with a horizontal scale of 1:2000 and a vertical scale of 1:159. The basin contains 5,000 cubic meters of water and covers 7,000 square meters. Three groups of pumps simulate tidal movement of water.

29. The National Trust was created in 1895 to limit the impact of industrialization and urbanization on the coastal zones of Great Britain. To date this foundation has acquired and protected nearly 1,120 kilometers of coastline.

30. The Conservatoire du Littoral is a public French establishment created in 1975. By 2003, it had acquired and protected 860 kilometers of the French coastline.

31. Genesis 9:8–17.

32. The astrolabe is one of the oldest navigation instruments, dating to 170 BC. It determines time and measures latitude. It was invented to measure a precise angle between the sun or the evening star and the horizon. With the aid of astronomical tables, the navigator can determine the hour and the latitude of his site. The astrolabe and the cross-staff are the ancestors of vertical and horizontal sextants and bearing circles.

33. The cross-staff, or Jacob's staff, was invented in the Middle Ages by Levi ben Gerson (1288–1344). It is composed of two mobile elements: the staff and the cross. It allows the measurement of the height of stars above the horizon by having the cross slide on the staff. This instrument allowed Vasco de Gama and Portuguese navigators to estimate latitude while at sea.

Epilogue

1. In 1939, the French illustrator Pierre Brissaud (1885–1964) produced a realistic reconstruction of the visit by the Queen of England to Van Leeuwenhoek's laboratory for the Abbott pharmaceutical laboratory. Van Leeuwenhoek is standing up with a microscope in each hand, lecturing Queen Mary II of England, who, in profile, examines something with another microscope. A black and white reproduction of this watercolor appears in Lisa Yount's book on Van Leeuwenhoek (Yount, *Antoni van Leeuwenhoek*).

2. de Lamarck, *Discours d'ouverture du cours de l'An VIII* (see chapter 1, note 14).

3. These citations are from the last pages of Darwin's great work, *On the Origin of Species* (484–85).

4. Gould, *Full House*.

5. Stephen J. Gould's book, *Full House: The Spread of Excellence from Plato to Darwin*, demystifies and debunks the notion of progress in evolution (p. 216).

6. The Environmental Commission of the Economic and Social Council of the Region Provence-Alps-French Riviera.

7. This was first coined by Charles of Valois-Bourgogne (Charles the Bold; 1433–1477), who ruled most of the territory of the United Provinces of The Netherlands. A century later, Prince William I of Orange (William the Silent; 1533–1584) adopted this motto. He headed the revolt against the Spanish that led to the independence of the United Provinces.

INDEX

The letter f *following a page number denotes a figure.*